The Humanitarian Emergency Settings Perceived Needs Scale (HESPER):

Manual with Scale

Acknowledgements

The development of the HESPER Scale was a collaborative project between the Department of Mental Health and Substance Abuse at the World Health Organization (WHO) Geneva, and the Institute of Psychiatry at King's College London (KCL).

Maya Semrau (KCL) developed the science of the HESPER Scale, conducted each of the seven pilot and field studies, and drafted this publication - all in close consultation with Dr Mark van Ommeren (WHO).

The Medical Research Council (UK) is gratefully acknowledged for funding Maya Semrau (KCL) through a three-year PhD studentship grant.

Steering Group:
The HESPER Project Steering Group consisted of Dr Mark van Ommeren and Dr Andre Griekspoor at WHO, and Professor Graham Thornicroft, Professor Louise M Howard, Dr Heidi Lempp, Dr Morven Leese and Maya Semrau (all at KCL).

International Advisory Group:
The HESPER International Advisory Group consisted of Dr Paul Bolton (John Hopkins University), Mr Kaz de Jong (Medicins Sans Frontieres Holland), Dr Nadine Ezard (Monash University), Prof Richard Garfield (Columbia University), Dr Johan Heffinck (at ECHO during the first two years of the project), Dr Lynne Jones (International Medical Corps), Dr Helen McColl (International Rehabilitation Council for Torture Victims), Dr Pau Pérez-Sales (Medicos del Mundo), Dr Shekhar Saxena (WHO), Dr Mike Slade (KCL), Dr Egbert Sondorp (London School of Hygiene and Tropical Medicine), Dr Zachary Steel (University of New South Wales), Dr Wietse Tol (HealthNet TPO, Yale University), and Dr Mike Wessells (Columbia University).

Gaza:
Data collection in Gaza was organized by Fafo Institute for Applied International Studies (Dr Åge Tiltnes and Mr Hani Eldada). The WHO Office in Gaza (Mr Dyaa Saymah) provided advice. Funding was provided by WHO Geneva and Fafo.

Haiti:
Data collection in Haiti was organized by International Medical Corps Haiti (Ms Isabelle Pilotte, Mr Daniel Joselito Charles, Mr Charles Lor and Mr Jason Erb).

Jordan:
Data collection in Jordan was organized by WHO Jordan (Dr Hashim El Mousaad, Dr Anita Marini, and Dr Nada Al Ward). Data collection was implemented by Accurate Opinion (field-testing), and the

Market Research Organisation (pilot-testing). Data collection was funded by the Jordanian Nursing Council, WHO Jordan, and the University of London Central Research Fund. UNHCR provided advice on sampling.

Nepal:
Data collection in Nepal was organized by HealthNet TPO / TPO Nepal (Mr Nagendra Luitel and Dr Mark Jordans). Funding was provided by WHO Geneva. UNHCR Nepal and WHO Nepal provided further support.

Sudan:
Data collection in Sudan was organized by Humanitarian Accountability Partnership (HAP International) (Ms Monica Blagescu).

United Kingdom:
Data collection in the United Kingdom was facilitated by the British Refugee Council (Ms Rachael Hardiman and Mr Alistair Griggs).

We thank all respondents, who participated in our 2008 expert survey, including Dr Alastair Ager (Columbia University), Mr F Jiovani Arias (Fundación Dos Mundos), Dr Nancy Baron (Global Psycho-Social Initiatives), Mr Mihir R Bhatt, (All India Disaster Mitigation Institute), Ms Christina Bitar (UNIFIL), Dr Cécile Bizouerne (Action Contre la Faim), Ms Nan Buzard (American Red Cross), Dr Jorge Castilla (WHO/PAHO), Prof Fatima Castillo (University of the Philippines), Dr Alessandro Colombo (International Rescue Committee), Ms Anjana Dayal (ICRC), Prof Joop de Jong (Vrije Universiteit Amsterdam), Dr Pamela DeLargy (UNFPA), Dr Linda Doull (Merlin), Dr Carolina Echeverri (WHO/PAHO), Dr Girma Ejere (Learning and Skills Council London), Dr Nadine Ezard (LSHTM), Mr Ananda Galappatti (Good Practice Group), Prof Rita Giacaman (Birzeit University), Dr Johan Heffinck (ECHO), Dr Lynne Jones (International Medical Corps), Dr Barbara Lopes Cardozo (Centers for Disease Control and Prevention), Dr Amanda Melville (UNICEF), Ms Carlinda Monteiro (Christian Children's Fund), Mr Charles Owusu (Christian Children's Fund), Ms Chrishara Paranawithana (WHO), Dr Jonathan Polonsky (WHO), Mr Bhava Poudyal (International Catholic Migration Commission), Dr Joe Prewitt (American Red Cross), Ms Sabine Rakotomalala (Terre des Hommes), Dr Bayard Roberts (LSHTM), Dr Jorge Rodriguez (WHO/PAHO), Prof Daya Somasundaram (University of Jaffna), Dr Peter Ventevogel (HealthNet TPO), Dr Johan von Schreeb (Karolinska Institute), Dr Vivien Walden (Oxfam GB), Dr Xiangdong Wang (WHO), Dr Mike Wessells (Columbia University), Ms Wendy Wheaton (independent consultant), Mr John Williamson (USAID), and Dr M Taghi Yasamy (WHO).

We thank Dr Xavier de Radigues (WHO) for reviewing the sections on sampling in this publication.

We thank all interviewers, translators, interpreters and other support staff in Gaza, Haiti, Jordan, Nepal, and Sudan.

Our special thanks go to the interviewed participants in Gaza, Haiti, Jordan (displaced Iraqi people), Nepal (Bhutanese refugees), Sudan, and the United Kingdom (refugees from the Democratic Republic of the Congo).

The Humanitarian Emergency Settings Perceived Needs Scale (HESPER):
Manual with Scale

Table of Contents

General Introduction 7

1 The HESPER Scale 8
   1.1 What is the HESPER Scale? – A brief overview 8
   1.2 Who may use the HESPER Scale? 8
   1.3 In what contexts may the HESPER Scale be used? 8
   1.4 How may the HESPER Scale be useful? 9
   1.5 Why was the HESPER Scale developed? 10
   1.6 What instrument is the HESPER Scale modelled on? 11
   1.7 How was the HESPER Scale developed? 12
   1.8 What are the HESPER Scale's psychometric properties? 16
   1.9 What is the final structure of the HESPER Scale? 16

2 The HESPER Assessment Process 17
   2.1 Overview of the HESPER assessment process 17
   2.2 The HESPER assessment process in detail 18
      2.2.1 Before interviews 18
         1.   Initial considerations 18
         2.   Deciding on your target population 18
         3.   Adapting the HESPER Scale and interviewers' training manual
            to the local context 19
         4a. Selecting respondents - Sampling 20
         4b. Selecting respondents - Sample size 22
         5.   Recruiting interviewers 23
         6.   Rapid training of interviewers 23
      2.2.2 During interviews – Issues to consider 24
         1.   Informed consent 24
         2.   Confidentiality 25
         3.   Standardizing interviews 25
         4.   Supervising interviewers 25
         5.   Minimizing non-response 25
         6.   Safety of interviewers and respondents 26
         7.   Self-care of interviewers 26

		2.2.3 During interviews - The interview process from the interviewers' perspective 26
	1.	An overview of the HESPER interview process 26
	2.	Making ratings on the HESPER Scale 27
		2.2.4 After interviews. 28
	1.	Data entry 28
	2.	Data analyses. 28
	3.	Presenting data 30
	4.	Identifying potential errors and biases in the results. 30
	5.	Communicating results to relevant stakeholders 31
	6.	Conducting follow-up in-depth assessments 31
	7.	Encouraging stakeholders to address prioritized needs 31
	8.	Monitoring perceived needs over time. 31

3 References 32

4 Appendices 37
Appendix 1 - Humanitarian Emergency Settings Perceived Needs Scale (HESPER) 38
Appendix 2 - HESPER Training Manual for Interviewers 41
Appendix 3 - Example HESPER Report 71
Appendix 4 - Sampling Guide 80
Appendix 5 - Kish Table 83
Appendix 6 - Performing Sample Size Calculations 84
Appendix 7 - Calculating Confidence Intervals. 87
Appendix 8 - Example Participant Information Sheet / Consent Form 89

Darfur, 2004, © WHO

General Introduction

Needs assessments are vital to identify the needs that are present in an affected population, and to inform the humanitarian response. There have been repeated recommendations for increased participation of affected populations in humanitarian assessment. Participation is seen as essential for avoiding basic mistakes in resource allocation, programme design, accountability, and for supporting psychosocial well-being.

In the humanitarian field, most needs assessments tend to use either population-based "objective" indicators (for example malnutrition or mortality indicators), or qualitative data based on convenience samples (for example through focus groups or key informant interviews). Whilst the latter method is not able to paint a full population-picture, the former is not able to gather information on people's subjective perception of needs.

The HESPER Scale was developed to fill this gap. It aims to provide a method for assessing perceived needs in representative samples of populations affected by large-scale humanitarian emergencies in a valid and reliable manner.

This manual includes the HESPER Scale (see Appendix 1), as well as a detailed explanation of how to use the HESPER Scale, how to train interviewers, and how to organise, analyze and report on a HESPER survey.

1. The HESPER Scale

1.1 WHAT IS THE HESPER SCALE? – A BRIEF OVERVIEW

The Humanitarian Emergency Settings Perceived Needs Scale (HESPER) (see Appendix 1) aims to provide a quick, scientifically robust way of assessing the *perceived serious needs* of people affected by large-scale humanitarian emergencies, such as war, conflict or major natural disaster. Perceived needs are needs which are felt or expressed by people themselves and are problem areas with which they would like help.

The HESPER Scale assesses a wide range of social, psychological and physical problem areas. However, it does not provide an answer as to whether, or how to, offer help. It simply aims to identify those serious perceived problems that are common in a population. These problems should then be assessed and addressed in more detail.

The HESPER Scale was developed by the World Health Organization and King's College London in order to fill several gaps in the humanitarian field. It enables needs assessments to be based directly on the views of people affected by humanitarian emergencies, and provides a more accurate picture of the serious problems with which the overall emergency-affected population wants help.

1.2 WHO MAY USE THE HESPER SCALE?

The HESPER Scale may be used by anybody in its current form for non-commercial purposes. Should you wish to make any modifications to the scale, or translate the scale into another language, you will need to get permission from WHO Press (for contact details, see inside cover page). Currently the HESPER Scale (i.e. Appendix 1 only) is available in English, French, Spanish, Arabic, Nepali, and French / Haitian Creole. *Word* files of the different HESPER Scale language versions are available upon request.

1.3 IN WHAT CONTEXTS MAY THE HESPER SCALE BE USED?

The HESPER Scale is applicable to a wide range of humanitarian settings, including those caused by natural events (such as earthquakes, floods, tsunamis, volcanoes, hurricanes, droughts or epidemics), as well as during war or other large-scale conflict. The scale can be used in acute or chronic humanitarian situations, urban or rural settings, and camp or community contexts. Whilst the scale is designed to be used in low- and middle-income countries as this is where most large-scale disasters occur, it may potentially also have value in large disasters in high-income countries (e.g. involving population displacements such as after Hurricane Katrina).

The scale is intended for administration to people in the general adult population, and has not been tested for use in people under 18 years of age.

1.4 HOW MAY THE HESPER SCALE BE USEFUL?

The HESPER Scale may be administered to representative samples, to estimate the presence of perceived needs in a population.

Some of the advantages of the HESPER Scale are outlined in Box 1.

> **Box 1**
>
> **The advantages of the HESPER Scale are**
> - It can be completed rapidly (between 15 to 30 minutes on average).
> - It can be easily self-learned and used on the basis of a self-training manual by local staff (without extensive use of trainers).
> - It is culturally applicable to a wide range of populations and settings in low- and middle-income countries.
> - It is usable in convenience samples very early on in emergencies, and can be used in representative samples at later stages of an emergency, thereby creating the possibility of tracking people's perceived needs over time.
> - It is consistent with the IASC *Guidelines on Mental Health and Psychosocial Support in Emergency Settings* (1), which includes a focus on perceived needs.
> - It is valid (meaning that it measures what it was intended to measure) and reliable (meaning that it provides consistent results across different raters and at different times).
> - In addition to its *core* items that assess almost universally occurring needs, *locally developed* items may also be added to account for needs that are specifically relevant to the local context.
> - It promotes increased accountability towards and participation of the affected population.
> - It assesses perceived needs across a broad range of problem areas.
> - It is freely available and easy to use.

The HESPER Scale is, as far as we are aware, the first scale which has been shown to measure people's perceived needs in a reliable and valid manner in representative samples (see section 1.8 for psychometric results from three field-sites). It thus combines the strengths of survey research (i.e. representative samples) with that of participatory methods (i.e. measuring *perceived* needs). We are not aware of any other brief multi-sectoral tool with tested reliability and validity that quantifies the prevalence and distribution of people's perceived needs in the general population in humanitarian settings. The scale complements, rather than replaces, existing rapid participatory assessment methodologies, which are currently the standard method of assessing perceived needs.

The HESPER Scale enables the rapid identification of broad problem areas with which the population is likely to want help. This information can then be determined for the population or subpopulation to identify perceived needs of the populations affected. Subsequent in-depth participatory assessments are then needed to understand the expressed needs, and to decide what exact interventions and supports would be helpful. It is possible to disaggregate the results and provide population profiles according to gender, age groups, ethnicity, or other relevant subpopulation groupings. The scale focuses on needs as perceived by the adult population, which may include concerns for their children.

By administering the HESPER Scale at multiple times, the scale may also be used to monitor the degree to which the humanitarian response is perceived by the affected people to be meeting their needs. The scale is therefore in line with the aim of increased accountability towards and participation of crisis-affected populations in assessments (2-5).

Although the HESPER Scale was developed for use in representative samples, the scale may also be used in convenience samples. This may be appropriate in some situations - such as the first few days or weeks of a large sudden-onset crisis - where representative sampling may not be possible. Whilst information can be collected more quickly and easily by using convenience samples, it should be noted that it is unlikely to be representative of the population at large.

The scale may also be used during service delivery to help individuals better. Indeed, the HESPER Scale can be a helpful tool for case management, which is a key element of social work and mental health care.

1.5 WHY WAS THE HESPER SCALE DEVELOPED?

The HESPER Scale was developed with the aim of filling at least four important gaps in the humanitarian field. First, humanitarian workers currently have some difficulties in conducting population-based psychosocial needs assessments. In the Inter-Agency Standing Committee's *Guidelines on Mental Health and Psychosocial Support in Emergency Settings* (1), mental health and psychosocial needs are seen to be diverse. Needs may be related to illness which predates the emergency (e.g. pre-existing alcohol dependence), they may relate to events which occurred during the emergency (e.g. earthquake, exposure to violence), or they may relate to the current emergency situation (e.g. sources of stress in a newly set-up camp). Needs which relate to people's current circumstances are influenced by aid in a range of humanitarian sectors. A person may thus experience trauma- or loss-induced psychological distress, but at the same time may for instance also suffer severely due to a perceived lack of security and experiencing psychosocial needs related to water and sanitation (e.g. if the available toilet facilities are in an insecure location, or in such state that they undermine people's experience of dignity). The IASC mental health and psychosocial framework is consistent with multi-sectoral assessment of *perceived* needs to identify people's sources of stress. The IASC Guidelines recommend participatory multi-sectoral needs assessments but do not answer the question of how to do population-based perceived needs assessments.

Second, current studies tend to focus mostly on the epidemiology of mental disorders in populations exposed to emergencies. A key question in the humanitarian field is the extent to which the distress or disorder within an affected population results from either events that have already occurred (i.e. trauma or loss), or those arising from the recovery environment (e.g. stressors in the current context) (6-8). A questionnaire measuring perceived needs gives researchers a tool to answer this key question and inform mental health and psychosocial support policy and practice.

Third, there are increasing calls to assess people's perceived needs, and to use perceived needs as key indicators for project design, monitoring and evaluation (1-5, 9). Similarly, in a recent research agenda priority ranking exercise for humanitarian settings, three of the ten most highly prioritized research questions included the participation of affected populations; the identification of affected populations' stressors was ranked as top priority (10). Currently, perceived needs are assessed mostly through rapid participatory assessments, which tend to involve gaining rich, qualitative data from selected stakeholders. Although very valuable, such rapid participatory assessments cannot provide a population picture. Currently, in the humanitarian field most population-based quantitative assessments are of "objective" indicators, such as mortality, nutrition and livelihood data. These indicators are often defined by outsiders (i.e. non-members of the affected population) and are - as far as we know - unable to quantify the prevalence and distribution of needs as perceived by members of the population themselves. The HESPER Scale may thus fill a gap by providing population-based quantitative assessments of perceived needs, based directly upon the views of those affected by the disaster.

Fourth, existing humanitarian needs assessment tools typically have unknown reliability or validity. For example, basic statistical knowledge on inter-rater reliability is essential in estimating the extent to which the results of assessments are likely to vary across interviewers. Similarly, test-retest statistics are necessary in order to know the extent to which interviewers gather consistent responses over time. Furthermore, information on criterion-related validity (i.e. strength of the relationship with a measurable external criterion) is helpful in judging whether a tool assesses what it purports to measure. The scientific study of reliability and validity of assessment (also called psychometrics) has, as far we are aware, never been applied to multi-sectoral humanitarian assessments. Psychometrics, a discipline originally developed by psychologists, is now widely applied in a range of disciplines (e.g. engineering, general medicine, health economics), whether or not the instruments measure underlying psychological constructs.

1.6 WHAT INSTRUMENT IS THE HESPER SCALE MODELLED ON?
The HESPER Scale focuses on diverse needs in the general population. It was modelled after a mental health instrument, the Camberwell Assessment of Need Short Appraisal Schedule (CANSAS) (11), which measures both met and unmet perceived needs of people with mental disorders across 22 domains of life. The CANSAS is a shortened version of the Camberwell Assessment of Need (CAN) with well-established reliability and validity (12, 13). The CANSAS is now the most widely used needs assessment instrument for people with mental health problems. It has been modified successfully for use in adults who have both mental disorders and learning disabilities (14), elderly people with

mental disorders (15), mothers with mental disorders (16), and forensic populations (17). The CAN has been translated into at least 25 languages, and has been adapted for use in several countries (18). Both the CAN and CANSAS have been used on a wide range of populations including asylum seekers and refugees in the UK (19, 20), as well as torture victims in centres of the International Rehabilitation Council for Torture Victims (IRCT) across several countries. The different versions of the CAN are available through www.iop.kcl.ac.uk/prism/can

1.7 HOW WAS THE HESPER SCALE DEVELOPED?

The HESPER Scale was developed over three phases:

- <u>Phase 1 (2008):</u> Development of a first draft scale through a process of item generation and item reduction, based on first a literature review, and second a survey with humanitarian experts.
- <u>Phase 2 (2009):</u> Pilot-testing of the draft scale in Jordan with displaced Iraqi people, in Gaza, Sudan, and in the UK with refugees from the Democratic Republic of the Congo (DRC), to assess the scale's feasibility, intelligibility and cultural applicability, and to establish the suitability of training materials.
- <u>Phase 3 (2010):</u> Field-testing of the revised draft scale in Jordan with displaced Iraqi people, in Haiti with people living in post-earthquake displacement camps, and in Nepal with Bhutanese refugees, to assess its psychometric properties (i.e. reliability and validity).

Phase 1: Development of the first draft scale

A zero draft and first draft of the scale (21) (available upon request) incorporating the universally relevant core need items were developed in 'Phase 1' of the project through a process of item generation and item reduction. For item generation a long list of potential need items for inclusion into the draft scale were extracted from relevant literature (grey and peer-reviewed). Only sources which directly dealt with emergency-affected people's views of *perceived* needs were employed, such as previous humanitarian needs assessments, existing NGO assessment reports, and published journal articles on perceived needs (see Table 1). Only items that were mentioned at least twice in any of these sources were included into the zero draft scale.

Table 1: Sources employed in the HESPER item generation process

Source	Country of study	Type of disaster	Period of data collection	Type of data collection
Asia				
Fritz Institute 2005 (22)	Indonesia, India, Sri Lanka	tsunami	Oct 2005	structured interviews
Fritz Institute 2007 (23)	Indonesia	earthquake	May – June 2006	face-to-face interviews with structured questionnaire
Fritz Institute 2007 (24)	Indonesia	tsunami	July 2006	face-to-face interviews with structured questionnaire
Poudyal et al 2007 (25)	Indonesia	conflict	Sept 2006	free-listing exercises, key informant interviews, focus groups
Thapa & Hauff unpublished (26)	Nepal	conflict	June – July 2003	cross-sectional household survey
Fritz Institute 2006 (27)	Pakistan	earthquake	Aug 2006	structured interviews
Africa				
Barton & Mutiti 1998 (28)	Uganda	conflict	Jan – Apr 1998	key informant interviews, focus groups
Betancourt et al 2009 (29)	Uganda	conflict	July – Aug 2004	free-listing exercises, key informant interviews
Bolton & Ndogoni 2000 (30)	Rwanda	conflict and genocide	Oct – Dec 1999	free-listing exercises, key informant interviews, pile sorts
Lee & Bolton 2007 (31)	Kenya	conflict	Oct – Nov 2005	free-listing exercises, key informant interviews
Murray et al 2006 (32)	Congo (DRC)	conflict	Feb 2006	free-listing exercises, key informant interviews
Briant & Kennedy 2004 (33)	Egypt	conflict	not known	interviews with questionnaires
Middle East				
Giacaman et al 2007 (34)	Occupied Palestinian Territories	conflict	Sept – Nov 2004	focus groups
Central America				
Pérez-Sales et al 2005 (35)	El Salvador	earthquakes	Apr 2001	semi-structured interview (CCI), including qualitative responses, focus groups

Need items were then selected and reduced into the first draft scale based on a survey with a wide range of purposively sampled general and psychosocial humanitarian experts across the world (24 male and 19 female), as well as six national aid workers in Sierra Leone. The survey included both quantitative and qualitative responses. For quantitative analyses participants rated the need items which had been compiled during the item generation stage of the project on an 11-point scale (0 to 10) of importance for inclusion into the scale, and suggested additional perceived need items that they considered important for inclusion. In addition, participants were encouraged to provide any further comments or feedback (21) (available upon request).

Since all items were rated as at least moderately important by participants, a broad approach was taken in the selection of items into the draft scale. The revision of items therefore primarily involved their rephrasing and regrouping. One item was added based on participants' suggestions. An overview of the changes that were made to need items based on this survey is available upon request (21).

A section was also introduced to the HESPER Scale whereby those needs which have been rated to be present by respondents are ranked in their order of importance, where numerous needs are unmet. This may enable prioritisation of needs and emergency relief to those areas where it is perceived to be needed most.

Subsequently, the draft HESPER Scale was reworded to make it more intelligible for respondents, and was restructured in terms of the order of its items (with basic physical survival needs first, and items covering community issues last). The rating scale was simplified to ease use of the scale in the field. An interviewers' training manual was also developed (see Appendix 2).

Phase 2: Pilot-testing

The draft HESPER Scale was then pilot-tested in Jordan with displaced Iraqi people, in Gaza, and in the South of Sudan, after having been pre-tested in the UK with refugees from Democratic Republic of the Congo (for details see Table 2). Pilot-testing was a learning exercise to understand the scale's feasibility, intelligibility and cultural applicability (cf. van Ommeren et al 1999 (36)), as well as assessing methodologies for subsequent field-testing. During pilot-testing respondents were:

- Administered the draft HESPER Scale by local interviewers familiar with the cultural setting. For a sub-sample, a silent rater was also present to assess inter-rater reliability. This involves two raters (one interviewer and one observer) making ratings during an interview with a respondent. It is then assessed how consistent their ratings are.
- Administered a participant survey, in which they were asked whether they thought that any perceived need items were missing from the draft scale and the extent to which the scale was understandable.
- Participated in focus group discussions, in which they were asked to report on the intelligibility, (cultural) acceptability, relevance and comprehensiveness of the scale's items. The suitability of the content and concepts were also checked.
- Interviewers were asked to complete an interviewer survey, in which they provided feedback

on the intelligibility of the HESPER Scale and interviewers' training manual, and whether they experienced any difficulties in conducting the interviews.

Following pilot-testing, the HESPER Scale was revised into a slightly shorter draft for field-testing (21) (available upon request), and revisions were made to the interviewers' training manual.

Table 2: HESPER pilot-testing and field-testing sites

Pilot- or field-testing	Month / Year	Country	Location	Population	Sample size	Local partner(s)
Pilot-testing	May 2009	United Kingdom	Hull	Refugees from Democratic Republic of the Congo, resettled from refugee camps in Zambia	7	British Refugee Council
Pilot-testing	June 2009	Jordan	Amman	Displaced persons from Iraq	40	WHO Jordan
Pilot-testing	October 2009	Occupied Palestinian Territories	Gaza City	Local population	40	Fafo Gaza, WHO Gaza
Pilot-testing	December 2009	Sudan	Juba	Local population	42	Humanitarian Accountability Partnership (HAP International)
Field-testing	July 2010	Jordan	Amman, Zarqa, Irbid, Madaba	Displaced persons from Iraq	269	WHO Jordan, Jordanian Nursing Council, UNHCR
Field-testing	September 2010	Haiti	Port-au-Prince, Jacmel	Local population in post-earthquake displacement camps	279	International Medical Corps Haiti
Field-testing	October / November 2010	Nepal	Beldangi-II camp, Jhapa district	Refugees from Bhutan	269	HealthNet TPO/ TPO Nepal, UNHCR Nepal, WHO Nepal

Phase 3: Field-testing

After having pilot-tested the scale in small samples, the HESPER Scale was then field-tested with larger samples. Field-testing took place in Jordan with displaced Iraqi people, in displacement camps in Haiti (eight months after the 2010 earthquake), and in Nepal with Bhutanese refugees (see Table

2). The goals for field-testing were to assess the HESPER Scale's reliability and validity:
- To measure inter-rater reliability, a second interviewer acted as silent rater.
- To assess test-retest reliability, respondents were interviewed a second time one week after the first interview by the same interviewer.
- Criterion (concurrent) validity is established by assessing the extent to which a new measure correlates with other existing similar measures administered at the same time. To test this, relevant HESPER items were compared to similar questions of an established quality-of-life instrument, the World Health Organization Quality of Life-100 Instrument (WHOQOL-100) (37).

A few small changes in the wording of seven items were made to finalize the HESPER Scale following field-testing.

1.8 WHAT ARE THE HESPER SCALE'S PSYCHOMETRIC PROPERTIES?

The results of field-testing were as follows (detailed results are available upon request):
- One-week test-retest reliability (intraclass correlation coefficients (ICCs); absolute agreement) – for the full scale – was 0.961 in Jordan, and 0.773 in Nepal (it should be noted that item-level test-retest reliability was low for some items in Nepal). Test-retest reliability was not assessed in Haiti.
- Inter-rater reliability ICCs (absolute agreement) were 0.998 in Jordan, 0.986 in Haiti, and 0.995 in Nepal.
- Correlations with selected items of the WHOQOL-100 were roughly as predicted, suggesting criterion (concurrent) validity in the three tested sites.
- In terms of face and content validity, survey respondents in 'Phase 1' of the project considered the list of HESPER items to be comprehensive and appropriate, and found each of the HESPER items to be of at least moderate importance on average. Focus group participants during pilot-testing in Jordan, Gaza and Sudan ('Phase 2') also considered the list of HESPER items to be comprehensive, and found all HESPER items to be understandable, relevant and culturally applicable.
- The average time to administer the HESPER Scale was 15 minutes (SD=4 min) in Jordan, 21 minutes (SD=12 min) in Haiti, and 22 minutes (SD=6 min) in Nepal.

1.9 WHAT IS THE FINAL STRUCTURE OF THE HESPER SCALE?

The final version of the HESPER Scale is displayed in Appendix 1. Perceived needs are assessed across 26 need items, which each include a short item heading, as well as an accompanying question. Ratings are then made for each need item according to unmet need (or serious problem, as perceived by the respondent), no need (or no serious problem, as perceived by the respondent), or no answer (i.e. not known, not applicable, or answer declined). Respondents are also asked to name any other unmet needs not already listed. Among items that have been rated as unmet need, respondents are asked to rank their three most serious problems (hereafter referred to as priority ratings).

2. The HESPER Assessment Process

2.1 OVERVIEW OF THE HESPER ASSESSMENT PROCESS

This section provides a brief overview of the HESPER assessment process. The remainder of this document then explains how to implement this process in more detail.

Please note that while this manual provides all the necessary information on the HESPER Scale and its use, the manual only provides a brief overview of the other processes involved in the design, implementation and analyses of a survey. Readers may need to consult other resources (see Reference section), or collaborate with experienced colleagues in order to carry out a HESPER survey.

Before interviews
1. Obtain necessary permissions to conduct the HESPER survey. Ensure that you have sufficient resources and time for the survey.
2. Decide on your target population.
3. Prepare the HESPER Scale (see Appendix 1) and interviewers' training manual (see Appendix 2) for use in the local context and target population.
4. Decide on your sampling method and sample size.
5. Recruit local interviewers to conduct the interviews.
6. Train interviewers to administer the HESPER Scale using the interviewers' training manual (see Appendix 2).

During interviews
1. Select and recruit respondents based on your sampling method.
2. Interview respondents.

After interviews
1. Enter data from the HESPER Scale into an electronic file. Data should be double-checked or double-entered to ensure accuracy, and then cleaned.
2. Analyse the data with a statistical programme. Possibly disaggregate the data according to sub-groups in the population.
3. Present the data in a table, graph or text format to obtain an overview of the results.
4. Identify potential errors and biases in the results.
5. Communicate the results to relevant stakeholders in plain language (see Appendix 3 for an example).
6. Follow-up the HESPER assessment with key informant interviews or focus group discussions, to obtain a more in-depth understanding of the results.
7. Encourage stakeholders to address prioritized needs.
8. Where possible, monitor changes in perceived needs over time.

2.2 THE HESPER ASSESSMENT PROCESS IN DETAIL

2.2.1 Before interviews

1. Initial considerations

You will likely need to get permission from the local or governmental authorities to conduct your survey. Where feasible and appropriate, obtain the support of the local community by informing them of your plans for the survey, the reasons for wishing to conduct the survey, and the methods used. Random sampling (see section 2.2.1 – 4a) for example can cause conflict and concern if members of the affected population do not understand why others are selected for interview and they are not. The results of the survey should be fed back to the relevant stakeholders (including the affected population) after completion.

Ensure that you have sufficient resources (i.e. funds, access to staff, transport etc.) before conducting the survey, and that the survey will be feasible (e.g. in terms of security, accessibility of populations, time required etc). See Box 2 for potential costs that may arise during a HESPER assessment.

Box 2

Potential costs of a HESPER assessment

- Interviewer and supervisor salaries
- Training costs, including:
  - Trainer's salary
  - Payment for interviewers' time during training
  - Food and drink for interviewers during training
  - Training materials, e.g. printing of interviewers' training manual, stationary etc.
- Payment of support staff, e.g. interpreter, data entry clerk, security staff etc
- Transport costs, e.g. to transport interviewers to interview locations
- Translation of materials before interviews (where applicable)
- Translation of data after interviews (where applicable)
- Printing of materials for interviews
- Other materials, e.g. stationary, ID cards for interviewers etc

---

2. Deciding on your target population

Examples of target populations to which the HESPER Scale may be applicable are people living in a particular displaced persons camp, those living in several displaced persons camps across a particular area, or the population of an entire village, city or country.

You may decide to employ inclusion or exclusion criteria. Please note that the HESPER Scale's psychometric properties have so far been tested in adults only.

3. Adapting the HESPER Scale and interviewers' training manual to the local context

As the HESPER Scale should be used in the local language, you will first need to find out whether the required language version already exists. If it does not, you will need to translate the HESPER Scale into the relevant language, which requires permission from WHO Press (see inside cover page for details). If possible, the interviewers' training manual (see Appendix 2) should also be translated into the language spoken by interviewers.

The HESPER Scale includes the term 'community' in several places. This term should be replaced with the term most suitable to the local geographical context (e.g. village, town, neighbourhood, camp etc) throughout the HESPER form before interviews take place.

When using the HESPER Scale, additional socio-demographic characteristics should always be collected, in order to be able to make comparisons between different sub-groups of the population. These should at the very least include gender and age. Other example variables which may be useful to collect include marital status, number of children, location (e.g. name of town, camp, or area of city), employment status, occupation, years of formal education, or length of time of displacement. It is also important to record the interviewers' name on each data sheet, as well as a participant number. Participant numbers should be used instead of names, to ensure confidentiality and anonymity of data.

One of the advantages of the HESPER Scale is that in addition to the core items, there is the option to add items specific to the local context. These context-specific items may be chosen based on previous key informant interviews or focus group discussions, or on field-observations of potentially important issues. Table 3 displays examples of additional items which may be relevant in some contexts. These are items which were added in some of the HESPER field-testing sites.

Darfur, 2004, © WHO

Table 3: Examples of context-specific HESPER items

Context-specific HESPER items	Relevant in what contexts?
Residency or resettlement Do you have a serious problem because you do not have residency where you live, or because you have not been resettled to another country?	particularly relevant in surveys with refugees
Burying and mourning the dead in your community Is there a serious problem in your community because bodies of the dead have not been dealt with according to people's religious and cultural beliefs?	particularly relevant when mortality is very high

4a. Selecting respondents - Sampling

To make up a representative sample, the selection of sampling units (i.e. participants or households) into the study needs to be random. The three most common probability (or random) sampling methods are:

- simple random sampling
- systematic random sampling
- cluster sampling

Simple random sampling and systematic random sampling involve simpler techniques than cluster sampling. However, in humanitarian settings they may often not be possible. A flowchart to determine which sampling method may be most suitable for a HESPER assessment is displayed in Figure 1.

Dacope, Bangladesh, following Cyclone Aila, 2011, © Sandie Walton-Ellery

Figure 1: Flowchart to determine which sampling method to use

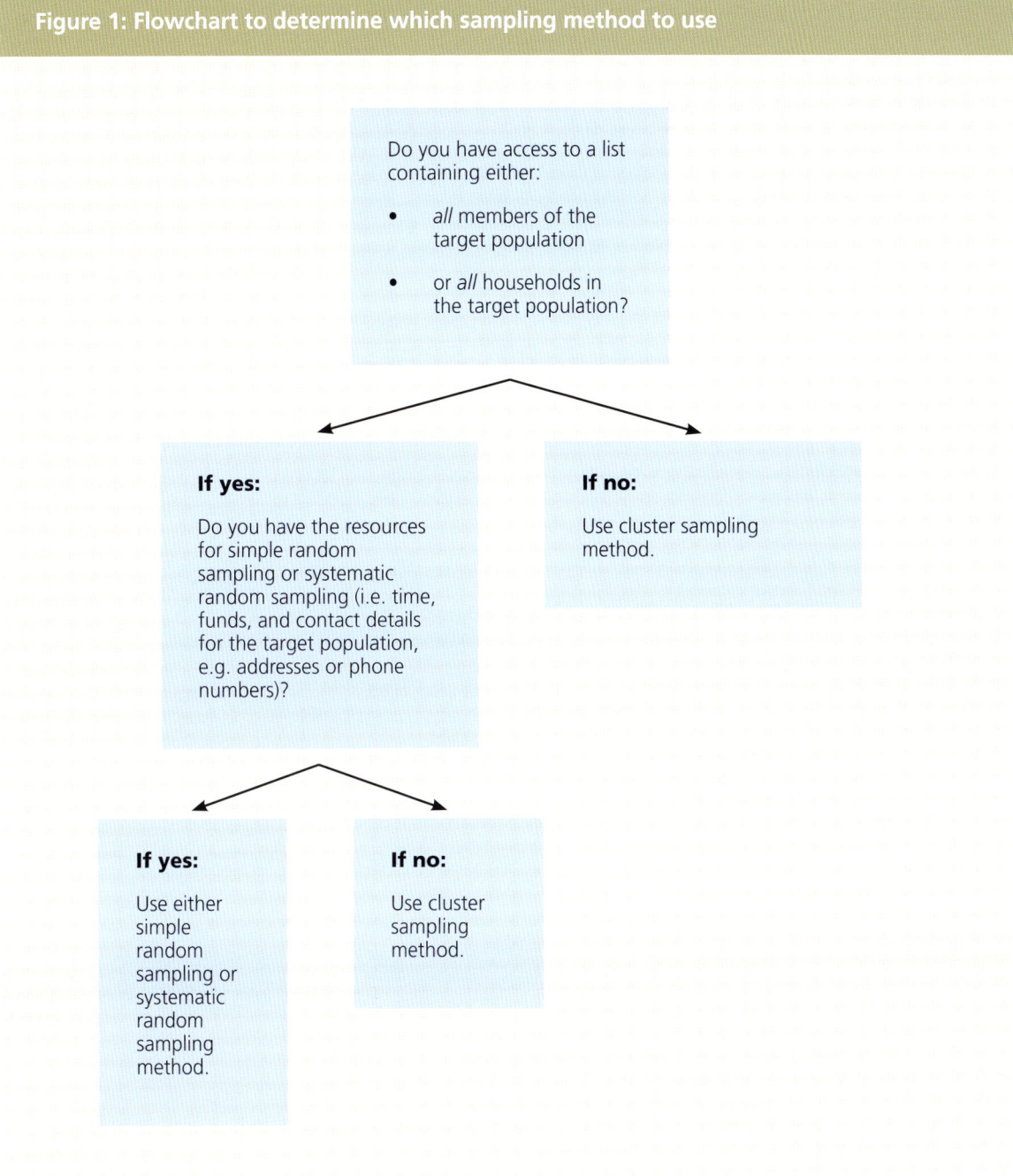

See Appendix 4 for a sampling guide, including step-by-step guides for simple random sampling, systematic random sampling, and cluster sampling. As cluster sampling is an advanced sampling technique, it is recommended to consult with an experienced epidemiologist or survey expert when applying this method.

4b. Selecting respondents - Sample size

A sample size calculation should be performed before a HESPER survey is conducted, to estimate the sample size needed to draw conclusions about the level of perceived needs in the population (i.e. to predict the frequency with which each of the HESPER Scale's items is perceived as serious problem (or no serious problem) in the wider target population).

There are many computer programs available which are able to perform such sample size calculations, some of which are accessible on the internet. If you are not familiar with such calculations, it is recommended to seek advice from an experienced statistician or epidemiologist.

Table 4 below gives suggested sample sizes for HESPER surveys. For justifications and further details of these sample size calculations (including worked examples), see Appendix 6.

Table 4: Required sample size for HESPER surveys (to accurately predict the frequency with which each HESPER item is perceived as serious problem in the population)*

	Expected response rate of 70%	Expected response rate of 80%	Expected response rate of 90%
Simple random sampling or systematic random sampling	137	120	107
Cluster sampling	274	240	213

* Assumptions: Level of precision (*beta*) = 0.1; Risk of error (*alpha*) = .05; Expected prevalence = 50%; Design effect (for cluster sampling) = 2

Should you wish to be able to make statistical comparisons between sub-groups, other calculations will need to be performed to estimate the required sample size. As with the sample size calculations above, sample size calculations for these analyses take into account the expected prevalence or mean, the level of precision, and a value related to the risk of error. In addition, information is needed on the number of sub-groups, as well as an estimate of whether there are likely to be an equal number of respondents in each of the sub-groups. It is likely that a much higher sample size will be required for these analyses.

5. Recruiting interviewers

Before data collection, you will need to recruit interviewers to administer the HESPER Scale. As with all assessments in humanitarian situations, interviewers selected to administer the scale should possess good interpersonal skills. They should have training and skills in basic interviewing and the application of relevant ethical principles, such as understanding the importance of confidentiality and informed consent. It is advisable for interviewers to have had an education of 12 years minimum (i.e. high school diploma or equivalent).

Furthermore, it is important that interviewers are familiar with the local setting within which the assessment is being conducted and that the choice of interviewer is suitable to the local culture. For example, in some cultures it may not be appropriate for a man to interview a woman, or for a younger woman to interview an older woman. Within the same country there may also be different cultural norms between particular groups, for example across different age groups, genders, or people of different religious beliefs. The choice of interviewer should therefore not only be appropriate to the overall population but also to the particular group. If working in another culture, interviewers should also ensure that their behaviour during the interview fits in with the cultural setting within which the interview is being conducted. This includes, for example, dressing according to the cultural norms and acting in a way that is locally acceptable.

Generally, the more interviewers you recruit, the faster data collection can be completed. However, if a large number of interviewers are recruited, it may make it more difficult to adequately train and supervise them. The number of interviewers recruited will therefore depend on the required sample size, the gender balance, as well as on the resources available, for instance the number of field supervisors, time, geographical spread of the sample, and funds. During field-testing of the HESPER Scale, between 6 and 12 interviewers were recruited in each of the three field-testing countries, who together were able to complete between 330 and 385 interviews per country in 12 working days (using 12 interviewers) to 22 working days (using 6 interviewers), including time needed to train interviewers in administering the HESPER Scale. In many humanitarian situations, it may be difficult to conduct more than 2 to 3 interviews per interviewer per working day.

6. Rapid training of interviewers

Interviewers should be trained to use the HESPER Scale by using the interviewers' training manual (see Appendix 2). The training should be conducted in the language spoken by interviewers. At least one whole day (ideally more) should be spent on training, with at least half of this being spent on practice interviews and role plays. Interviewers should also be given sufficient time to read the interviewers' training manual, either during training or in their own time.

In addition to the training session, a pilot-test of at least half a day should be conducted, in which interviewers practice using the HESPER Scale in the field (e.g. interviewing members of the target population). These interviews should be observed by a knowledgeable supervisor. Following the pilot-trial, the supervisor should discuss the interview process with the interviewers, and discuss any problems with them.

During training, it may be useful to inform interviewers of any support structures or services available to respondents. At the end of interviews, they may then pass this information on to those respondents, who seem very upset or distressed by their situation (see section 3.5 in interviewers' training manual).

2.2.2 During interviews – Issues to consider

1. Informed consent

Administration of the HESPER Scale by interviewers to respondents should be preceded by an informed consent process. This is to ensure that respondents take part in the interview voluntarily, without coercion or fear that they will miss out on benefits if they do not participate, and to facilitate that no unrealistic expectations are raised.

Informed consent may be taken either verbally or in writing, depending on the context. At a minimum this should involve explaining to the respondent who the interviewer is and the agency he or she represents, the reasons for the survey, and an overview of the interview process, including the amount of time needed. Furthermore, it should be clarified that participation is anonymous, completely voluntary, that no compensation will be paid, and that there will be no benefits to respondents if they participate. The interviewer should then answer any questions the respondent may have, before asking whether the respondent agrees to take part.

Ideally each respondent should be given a participant information sheet explaining all of the above (which they may either read themselves, or which may be read out to them), and each respondent should sign two copies of this sheet (one for the respondent to keep, one for the interviewer) as consent to take part in the survey. If the respondent does not agree to take part, he or she should not be pressured into doing so. Respondents should also have the right to withdraw from the interview at any point without having to give a reason. See Appendix 8 for an example of a participant information sheet / consent form.

Pinchinat displacement camp, Jacmel, Haiti, 2010, © Maya Semrau

2. Confidentiality

In order to respect respondents' right to privacy, it is important that all their details and responses are kept confidential. This means that respondents' answers or personal details should not be discussed with other people outside of the assessment team. Members of the team should not discuss anything with others, even once the assessment has been completed. Furthermore, all individual data sheets should be kept confidentially, and no information from which the identity of respondents may be identified should be made public. Instead of respondents' names, pre-assigned numbers should be used on data sheets. Any information linking respondents' names with their numbers should be kept separately.

The interview should be conducted in a place which is as private as possible. Ideally this means that the interview should be conducted in a quiet room with only the interviewer and the respondent present. However, this may not always be possible or culturally appropriate.

3. Standardizing interviews

In order for results to be reliable, it is important that interviews are conducted in the same way for each respondent and across different interviewers. Interviewers should be given an equal amount of training, and should be given sufficient time to practice HESPER interviews. Standardizing interviews in this way ensures that any differences in results are not due to differences in the interview process, but are rather due to 'true' differences between respondents.

4. Supervising interviewers

Throughout data collection, interviewers should be supervised by a knowledgeable team leader who has experience of conducting surveys in the field. Supervisors should ideally meet with interviewers at the end of each day, and at least every few days during data collection, to review the interview process, discuss any problems, and to collect data sheets from interviewers.

5. Minimizing non-response

Even though there is usually some non-response in all surveys, non-response should be reduced as much as possible, as a low response rate may bias results. This is because there may be systematic differences between those who choose to participate in a survey, and those who do not (38).

To minimize non-response, if the potential respondent is not in, it may be useful to ask neighbours whether the dwelling is inhabited, and if so, at what time the residents tend to be home. If the dwelling is inhabited, or if this is unknown, two or more visits should be made to establish contact with the residents (38).

If a potential respondent declines to take part, they should not be pressured into taking part.

6. Safety of interviewers and respondents

It is important that both interviewers and respondents remain sufficiently safe throughout the interview and feel comfortable about the setting within which the interview takes place. Interviewers should therefore choose a setting which is safe and culturally appropriate. For example, it may sometimes not be suitable to conduct the interview in respondents' houses or shelters. In this case other arrangements should be made for the interview to take place in a quiet and suitable place. In some situations, it may not be appropriate or safe for women to be interviewed by a male interviewer, or vice versa.

Supervisors or project leaders should always make sure that somebody knows where and when each interview is taking place. If possible, both supervisors and interviewers should carry a mobile phone or, where appropriate, satellite phone with them. Depending on the setting, it may be necessary or advisable to do the interviews in pairs or for interviewers to have an escort with them.

7. Self-care of interviewers

It is possible that interviewers may feel upset or distressed by an interview, or that they find the interview process difficult. Supervisors or project leaders should invite interviewers to speak to them if this is the case or, if available, a staff welfare officer.

2.2.3 During interviews - The interview process from the interviewers' perspective

1. An overview of the HESPER interview process

Each interview should take around 15 to 30 minutes, but this will vary.

HESPER interviews can be summarised into six steps, which are outlined in Box 3.

The HESPER interview process is described in detail in section 2.2 of the interviewers' training manual (see Appendix 2).

Darfur, 2004, © WHO

Box 3

The HESPER interview in six steps from the interviewers' perspective

1. <u>Before the interview:</u> Make sure you are familiar with the HESPER Scale and its rating system. You should have practiced this with your colleagues before your first HESPER assessment.

2. <u>Introduction to the interview:</u> Introduce yourself to the person you are interviewing, explain the purpose of the interview and the interview process, answer any questions they may have, and ask if they agree to take part. If they do agree to take part, make sure that they are comfortable and ready to start the interview. Then write down the date, your name, the participant number, the location in which the person lives, as well as the person's gender and age at the top of the HESPER form.

3. <u>HESPER Scale – Need ratings:</u> Read out the text at the top of the HESPER form. Then ask questions about each of the HESPER Scale's problem areas and give each question a rating based on the person's answers. Write the ratings in the appropriate column as you go along. Ask one or more follow-up questions for each area if necessary to make sure that you understand the person's views correctly.

4. <u>HESPER Scale – Other serious problems:</u> Once you have rated each of the HESPER Scale's problem areas, ask the person whether they have any other serious problems and write these down in the assigned spaces at the bottom of the HESPER form.

5. <u>HESPER Scale – Priority ratings for serious problems:</u> Then ask the person to tell you their three most serious problems in order of importance and write these down in the assigned spaces at the bottom of the HESPER form.

6. <u>End of interview:</u> Thank the person for taking part in the interview, answer any questions they have, and make sure that they have your or your organisation's contact details.

2. Making ratings on the HESPER Scale

A HESPER interview involves asking respondents about 26 problem areas. Interviewers rate whether the person feels that they have a serious problem in that particular area based on the person's answers.

Each question is rated in the same way. The interviewer asks respondents about each problem area and makes a rating based on their answers. See Box 4 for an explanation of the HESPER Scale's rating system.

Box 4

Each question is rated according to the following guidelines

Rate 9 (does not know / not applicable / declines to answer) if the person does not know how to answer the question, does not want to answer the question, or if the question does not apply to them.

Rate 1 (serious problem) if the person thinks that there is a serious problem for this question. A serious problem is a problem which the person feels is serious (however they define this).

Rate 0 (no serious problem) if the person does not think that there is a serious problem for this question.

2.2.4 After interviews

1. Data entry

Data should be entered and analysed in a statistics programme. Data should at a minimum be double-checked, and ideally should be entered twice and then both data sets cross-checked, to ensure that the data have been entered correctly. If this is not feasible, then a randomly selected percentage (e.g. 20%) of the data should be re-entered or double-checked, to ensure that the data has been accurately entered.

The data should then be assessed for their quality by checking for *outliers*, *inconsistencies*, and *missing data*. An *outlier* is a data point, which seemingly does not fit in with the remainder of the data set, that is it lies outside the range of all other data. For instance, if all respondents rated between 2 and 10 of the HESPER items as serious problem, but one respondent rated all 26 items as serious problem, this data point would be an outlier.

Inconsistencies in the data may arise where two data points for one respondent are not compatible with each other. An example of this may be that one of the HESPER items was not rated as serious problem by the respondent, but was then mistakenly coded as a priority rating (i.e. was rated as one of the respondent's three most serious problems). Another example may be that a respondent apparently has two children, even though he or she is not married (in cultures where this is a taboo). When dealing with outliers or inconsistencies in the data set, re-check the electronic data against the original data sheets or re-interview the person, to ensure that the data are correct.

There are various ways to deal with *missing data* (some more complex than others). One of the simplest approaches (if the number of missing values is low, under around 20%, and there does not appear to be any patterns in the missing data) is to impute values by the taking the average value across other respondents. Other more complex methods should be applied if the number of missing values is high (over around 20%), or there appears to be a pattern in the missing data (e.g. a particular item is missing more than others, or one group of respondents has more missing items than another group). Should this be the case, it is recommended to seek advice from an experienced statistician or epidemiologist.

2. Data analyses

Quantitative statistical methods are required to derive population-based results for HESPER surveys. All statistical analyses can be performed by standard statistics programmes.

<u>Need ratings of individual HESPER items</u>

The formula to calculate prevalence (P) of need ratings for individual HESPER items, expressed as percentage, is:

P (%) = number of respondents who rated the HESPER item as serious problem (or alternatively no serious problem) / number of respondents interviewed x 100

Other serious problems

To derive results for any other serious problems (i.e. needs) named by respondents, these should be listed. Where appropriate, similar items may be grouped together. Prevalence of reported other serious problems may then be calculated according to the formula above at the bottom of page 28.

Priority ratings for serious problems

The formula to calculate prevalence (P) of priority ratings for individual HESPER items, expressed as percentage, is:

P (%) = number of respondents who rated the HESPER item as one of their three most serious problems / number of respondents interviewed x 100

or (where priority ratings are separated in the analyses):

P (%) = number of people who rated the HESPER item as either their first, second, or third most serious problem / number of respondents interviewed x 100

Total number of needs

In addition to being able to calculate the prevalence of individual problem areas in the sample, the HESPER Scale also allows the average total number of needs (or serious problems) that respondents have to be calculated. As long as data are normally distributed, the mean is the most appropriate measure of the average. For data to be normally distributed, they should form a bell-shaped curve when being displayed as histogram, and the mean, median, and mode (all measures of the average) should have roughly the same (or similar) values. Means should always be displayed together with their standard deviation.

If data are not normally distributed, then the median may be a better measure of the average. The median is the number which divides the set of numbers equally into a lower and an upper half, i.e. it is the half-way point of all numbers. The median is usually presented together with the range (i.e. the range between the minimum and maximum numbers).

Confidence intervals

Confidence intervals may be calculated for both the prevalence (i.e. percentages) of need ratings and priority ratings (for individual HESPER items), as well as for averages (e.g. means) of total number of needs. Though it is not necessary to calculate confidence intervals, they can be a useful way to communicate results. See Appendix 7 for the formulae to calculate confidence intervals.

Disaggregating results according to sub-populations

To disaggregate results according to sub-populations (e.g. different genders, different age groups, people living in different geographical areas etc.), the sample may be divided according to the particular sub-population, and then the analyses performed for each sub-population in the same way as for the entire sample. For example, if comparing results between men and women, either prevalence of need ratings or priority ratings, or the average (e.g. mean) total number of needs, may be calculated separately for men and women.

Statistical tests may be used to calculate whether any differences in ratings between groups are statistically significant. These include chi-squared tests (for categorical data, such as prevalence), or independent t-tests or ANOVA (for continuous data, such as means). Standard statistics programmes are able to perform these tests.

3. Presenting data

Tables and graphs may be useful for displaying HESPER data. Graphs which may be useful include pie charts or bar charts (for need ratings or priority ratings of individual HESPER items), or histograms (for total number of needs). Standard statistics programmes are able to create these (see Appendix 3 for an example of how to present data in a report of a HESPER assessment).

4. Identifying potential errors and biases in the results

As HESPER surveys provide population-level data, they may give a good indication of the prevalence of need, and the types of needs, present in a population. However, as with all surveys, results should be interpreted cautiously, as there is always the possibility of results being somewhat compromised by errors. Possible reasons for error should therefore be considered carefully. Generally, there are two types of error: random error and systematic error (i.e. bias). It is important to discuss relevant errors and biases in any report on a HESPER assessment.

Random error is "when a value of the sample measurement diverges – due to chance alone – from that of the true population value" (Bonita et al 2006, p. 52 (39)). In other words, the results are not representative of the population at large because of some chance factor. Two common types of random error are (39):

- Sampling error: This, as well as error due to variations between different respondents, may have occurred when the sample was too small (see section 2.2.1 – 4b).
- Measurement error: This may have occurred if the measures used were inaccurate in a non-systematic way. As the HESPER Scale has good psychometric properties (i.e. it has been shown to be reliable and valid in three field-settings), this goes some way in reducing measurement error. Yet, if the HESPER Scale is used in a new setting without a local study verifying its psychometric properties, then the possibility of measurement error should be acknowledged.

Systematic error occurs when "results differ in a systematic manner from the true values" (Bonita et al 2006, p. 53 (39)). In other words, the results are not representative of the population at large because of some factor in the sample which systematically deviates from that in the population at large. The two main types of systematic error are (39):

- Selection bias: This may have occurred if not all people in the target population had an equal chance of being selected into the study, for instance because non-probability sampling methods were used or because some mistake was made during sampling (see section 2.2.1 – 4a).
- Measurement bias: This may have occurred where the measurements used were systematically inaccurate. One type of measurement bias results from response sets where the respondent believes that it is better to over-estimate or under-estimate the seriousness of their needs, for example because respondents have raised expectations of the help they will receive (see section 3.3 in interviewers' training manual). Triangulating results with other methods (e.g. in-depth key informant interviews) may be necessary to identify such biases. Another type of measurement bias can result from observer bias, in which the interviewer influences the ratings made. For example, if interviewers believed that respondents over-estimated their serious problems, this may have influenced the way they made ratings. This may have occurred if interviewers were trained insufficiently (see section 2.2.1 – 6).

5. Communicating results to relevant stakeholders

The results of any HESPER assessment should be communicated to all relevant stakeholders in plain language. This may be done through a report which is disseminated to stakeholder groups (including members of the affected population). An example of such a report can be found in Appendix 3.

6. Conducting follow-up in-depth assessments

The HESPER Scale on its own is not sufficient to fully understand people's perceived needs. HESPER surveys should therefore be followed-up and triangulated with in-depth key informant interviews (e.g. with community leaders, traditional and religious healers, humanitarian workers), in-depth interviews with the affected population, observation, mapping exercises, or focus group discussions, to better understand the specifics of why – from the respondents' perspectives – needs are rated as they are.

7. Encouraging stakeholders to address prioritized needs

These in-depth qualitative assessments following HESPER surveys may be used to identify and develop suitable interventions to address needs as perceived by the affected population. Stakeholders should be encouraged to address these. Appropriate questions should be asked during key informant interviews or focus group discussions, to identify what resources and interventions are considered suitable or useful from the perspective of the affected population.

8. Monitoring perceived needs over time

Use of the HESPER Scale at one time point is not sufficient to understand the complexities of population needs. Needs assessments should be viewed and contextualised within the specific timeframe within which they are conducted; for this the HESPER Scale may be used repeatedly over time to identify shifts and trends in perceived needs and to assess whether prioritized needs are addressed over time.

3. References

1. Inter-Agency Standing Committee (IASC). IASC Guidelines on Mental Health and Psychosocial Support in Emergency Settings. Geneva: IASC; 2007.

2. Active Learning Network for Accountability and Performance in Humanitarian Action (ALNAP). Participation of Crisis-Affected Populations in Humanitarian Action: A Handbook for Practitioners. ALNAP; 2003.

3. Humanitarian Accountability Partnership (HAP International). The 2010 HAP Standard in Accountability and Quality Management. Geneva: HAP International; 2010.

4. Humanitarian Accountability Partnership (HAP International). HAP 2007 Standard in Humanitarian Accountability and Quality Management. Geneva: HAP International; 2007.

5. The Sphere Project. Humanitarian Charter and Minimum Standards in Disaster Response. Geneva: The Sphere Project; 2011.

6. Miller KE, Rasmussen A. War Exposure, Daily Stressors, and Mental Health in Conflict and Post-Conflict Settings: Bridging the Divide Between Trauma-Focused and Psychosocial Frameworks. Soc Sci Med. 2010;70:7-16.

7. Steel Z, Chey T, Silove D, Marnane C, Bryant RA, van Ommeren M. Association of Torture and Other Potentially Traumatic Events with Mental Health Outcomes Among Populations Exposed to Mass Conflict and Displacement. JAMA. 2009;302(5):537-49.

8. Porter M, Haslam N. Predisplacement and Postdisplacement Factors Associated with Mental Health of Refugees and Internally Displaced Persons: A Meta-Analysis. JAMA. 2005;294(5):602-12.

9. Emergency Capacity Building Project Impact Measurement and Accountability in Emergencies - The Good Enough Guide: Oxfam GB; 2007.

10. Tol WA, Patel V, Tomlinson M, Baingana F, Galappatti A, Panter-Brick C, Silove D, Sondorp E, Wessells M, van Ommeren M. Research Priorities for Mental Health and Psychosocial Support in Humanitarian Settings. submitted manuscript

11. Slade M, Thornicroft G, Loftus L, Phelan M, Wykes T. CAN: Camberwell Assessment of Need - A Comprehensive Needs Assessment Tool for People with Severe Mental Illness. London: Gaskell; 1999.

References

12. Phelan M, Slade M, Thornicroft G, Dunn G, Holloway F, Wykes T, et al. The Camberwell Assessment of Need: The Validity and Reliability of an Instrument to Assess the Needs of People with Severe Mental Illness. Br J Psychiatry. 1995;167:589-95.

13. Andresen R, Caputi P, Oades LG. Interrater Reliability of the Camberwell Assessment of Need Short Appraisal Schedule. Aust N Z J Psychiatry. 2000;34: 856-61.

14. Xenitidis K, Thornicroft G, Leese M, Slade M, Fotiadou M, Philp H, et al. Reliability and Validity of the CANDID - A Needs Assessment Instrument for Adults with Learning Disabilities and Mental Health Problems. Br J Psychiatry. 2000;176:473-8.

15. Reynolds T, Thornicroft G, Abas M, Woods B, Hoe J, Leese M, et al. Camberwell Assessment of Need for the Elderly (CANE) - Development, Validity and Reliability. Br J Psychiatry. 2000;176:444-52.

16. Howard L, Hunt K, Slade M, O'Keane V, Senevirante T, Leese M, et al. Assessing the Needs of Pregnant Women and Mothers with Severe Mental Illness: The Psychometric Properties of the Camberwell Assessment of Need - Mothers (CAN-M). International Journal of Methods in Psychiatric Research. 2007;16(4): 177-85.

17. Thomas S, Harty MA, Parrott J, McCrone P, Slade M, Thornicroft G. CANFOR: Camberwell Assessment of Need - Forensic Version. A Needs Assessment for Forensic Mental Health Service Users. London: Gaskell; 2003.

18. McCrone P, Leese M, Thornicroft G, Schene AH, Knudsen HC, Vazquez-Barquero JL, et al. Reliability of the Camberwell Assessment of Need--European Version. EPSILON Study 6. European Psychiatric Services: Inputs Linked to Outcome Domains and Needs. British Journal of Psychiatry - Supplementum. 2000(39):s34-40.

19. McColl H, Johnson S. Characteristics and Needs of Asylum Seekers and Refugees in Contact with London Community Mental Health Teams: A Descriptive Investigation. Soc Psychiatry Psychiatr Epidemiol. 2006 Oct;41(10):789-95.

20. McCrone P, Bhui K, Craig T, Mohamud S, Warfa N, Stansfeld SA, et al. Mental Health Needs, Service Use and Costs Among Somali Refugees in the UK. Acta Psychiatr Scand. 2005;111(5):351-7.

21. Semrau M. The HESPER (Humanitarian Emergency Settings Perceived Needs) Scale - Development of a Draft Instrument. [MSc Dissertation]. unpublished manuscript.

22. Fritz Institute. Recipient Perceptions of Aid Effectiveness: Rescue, Relief and Rehabilitation in Tsunami Affected Indonesia, India and Sri Lanka: www.fritzinstitute.org; 2005.

23. Fritz Institute. Recovering from the Java Earthquake: Perceptions of the Affected: www.fritzinstitute.org; 2007.

24. Fritz Institute. The Immediate Response to the Java Tsunami: Perceptions of the Affected: www.fritzinstitute.org; 2007.

25. Poudyal B, Erni T, Jonathan A, Subyantoro T, Saraswati IA, Hutapea M. Qualitative Assessment of Violence Affected Populations in Bireuen, Aceh, Indonesia: International Catholic Migration Commission (ICMC); 2007.

26. Thapa SB, Hauff E. Needs Assessment and Disability Among Displaced Persons During an Armed Conflict in Nepal. unpublished.

27. Fritz Institute. Surviving the Pakistan Earthquake: Perceptions of the Affected One Year Later: www.fritzinstitute.org; 2006.

28. Barton T, Mutiti A, The Assessment Team for Psycho-Social Programmes in Northern Uganda. NUPSNA - Northern Uganda Psycho-Social Needs Assessment. Uganda: UNICEF / UGANDA Government Publication; 1998.

29. Betancourt TS, Speelman L, Onyango G, Bolton P. A Qualitative Study of Mental Health Problems Among Children Displaced by War in Northern Uganda. Transcultural Psychiatry. 2009;46(2):238-56.

30. Bolton P, Ndogoni L. Cross-Cultural Assessment of Trauma-Related Mental Illness: CERTI Crisis and Transition Toolkit; 2000.

31. Lee K, Bolton P. Impact of FilmAid Programs in Kakuma, Kenya - Final Report: http://filmaid.org/where/bu%20report%20final%20evaluation.pdf; 2007.

32. Murray L, Bass J, Bolton P. Qualitative Study to Identify Indicators of Psychosocial Problems and Functional Impairment among Residents of Sange District, South Kivu, Eastern DRC: http://sph.bu.edu/images/stories/scfiles/cih/drc_qualitative_study_report_10-4-06-final.pdf. 2006.

33. Briant N, Kennedy A. An Investigation of the Perceived Needs and Priorities Held by African Refugees in an Urban Setting in a First Country of Asylum. Journal of Refugee Studies. 2004;17(4):437-59.

34. Giacaman R, Mataria A, Nguyen-Gillham V, Safieh RA, Stefanini A, Chatterji S. Quality of Life in the Palestinian Context: An Inquiry in War-Like Conditions. Health Policy. 2007;81(1):68-84.

35. Pérez-Sales P, Cervellón P, Vázquez C, Vidales D, Gaborit M. Post-Traumatic Factors and Resilience: The Role of Shelter Management and Survivours' Attitudes after the Earthquakes in El Salvador (2001). Journal of Community & Applied Social Psychology. 2005;15(5):368-82.

36. van Ommeren M, Sharma B, Thapa S, Makaju R, Prasain D, Bhattarai R, et al. Preparing Instruments for Transcultural Research: Use of the Translation Monitoring Form with Nepali-Speaking Bhutanese Refugees. Transcultural Psychiatry. 1999;36(3):285-301.

References

37. World Health Organization (WHO), The WHOQOL-Group, WHOQOL-100. Geneva: WHO; 1995.

38. Magnani R. Sampling Guide: Food and Nutrition Technical Assistance Project (FANTA); 1997.

39. Bonita R, Beaglehole R, Kjellström T. Basic Epidemiology, 2nd edition. Geneva: World Health Organization; 2006.

40. Brennan M. 'Epidemiology' Course. Trinity College Dublin. December 2009.

41. Kish L. A Procedure for Objective Respondent Selection Within the Household. J Amer Statistical Assoc. 1949;44(247):380-7.

Appendices

Appendix 1   **Humanitarian Emergency Settings Perceived Needs Scale (HESPER)**

Appendix 2   **HESPER Training Manual for Interviewers**

Appendix 3   **Example HESPER Report**

Appendix 4   **Sampling Guide**

Appendix 5   **Kish Table**

Appendix 6   **Performing Sample Size Calculations**

Appendix 7   **Calculating Confidence Intervals**

Appendix 8   **Example Participant Information Sheet / Consent Form**

Appendix 1 - Humanitarian Emergency Settings Perceived Needs Scale (HESPER)

Date:	Interviewer name:	Participant number:
Location (name of city, village or camp):	Gender:	Age:

Rating: 0 = no serious problem 1 = serious problem 9 = does not know / not applicable / declines to answer	Ratings

I am going to ask you about the **serious problems** that you may **currently** be experiencing. We are interested in finding out what you think – a serious problem is a problem that **you** consider serious. There are no right or wrong answers. I am going to ask you about your own serious problems first.

1. Drinking water Do you have a serious problem because you do not have enough water that is safe for drinking or cooking?	
2. Food Do you have a serious problem with food? For example, because you do not have enough food, or good enough food, or because you are not able to cook food.	
3. Place to live in Do you have a serious problem because you do not have a suitable place to live in?	
4. Toilets Do you have a serious problem because you do not have easy and safe access to a clean toilet?	
5. Keeping clean *For men:* Do you have a serious problem because in your situation it is difficult to keep clean? For example, because you do not have enough soap, water or a suitable place to wash. *For women:* Do you have a serious problem because in your situation it is difficult to keep clean? For example, because you do not have enough soap, sanitary materials, water or a suitable place to wash.	
6. Clothes, shoes, bedding or blankets Do you have a serious problem because you do not have enough, or good enough, clothes, shoes, bedding or blankets?	
7. Income or livelihood Do you have a serious problem because you do not have enough income, money or resources to live?	
8. Physical health Do you have a serious problem with your physical health? For example, because you have a physical illness, injury or disability.	
9. Health care *For men:* Do you have a serious problem because you are not able to get adequate health care for yourself? For example, treatment or medicines. *For women:* Do you have a serious problem because you are not able to get adequate health care for yourself? For example, treatment or medicines, or health care during pregnancy or childbirth.	
10. Distress Do you have a serious problem because you feel very distressed? For example, very upset, sad, worried, scared, or angry.	
11. Safety Do you have a serious problem because you or your family are not safe or protected where you live now? For example, because of conflict, violence or crime in your community, city or village.	
12. Education for your children Do you have a serious problem because your children are not in school, or are not getting a good enough education?	
13. Care for family members Do you have a serious problem because in your situation it is difficult to care for family members who live with you? For example, young children in your family, or family members who are elderly, physically or mentally ill, or disabled.	
14. Support from others Do you have a serious problem because you are not getting enough support from people in your community? For example, emotional support or practical help.	
15. Separation from family members Do you have a serious problem because you are separated from family members?	
16. Being displaced from home Do you have a serious problem because you have been displaced from your home country, city or village?	

Source: World Health Organization & King's College London (2011). *The Humanitarian Emergency Settings Perceived Needs Scale (HESPER): Manual with Scale*. Geneva: World Health Organization. Requests for permission to reproduce, adapt or translate this scale should be addressed to WHO Press through the WHO web site (http://www.who.int/about/licensing/copyright_form/en/index.html).

Interviewers should be trained in the HESPER before use (see Appendix 2 of the HESPER manual).

17. Information *For displaced people:* Do you have a serious problem because you do not have enough information? For example, because you do not have enough information about the aid that is available; or because you do not have enough information about what is happening in your home country or home town. *For non-displaced people:* Do you have a serious problem because you do not have enough information? For example, because you do not have enough information about the aid that is available.	
18. The way aid is provided Do you have a serious problem because of inadequate aid? For example, because you do not have fair access to the aid that is available, or because aid agencies are working on their own without involvement from people in your community.	
19. Respect Do you have a serious problem because you do not feel respected or you feel humiliated? For example, because of the situation you are living in, or because of the way people treat you.	
20. Moving between places Do you have a serious problem because you are not able to move between places? For example, going to another village or town.	
21. Too much free time Do you have a serious problem because you have too much free time in the day?	

The last few questions refer to people in your community*, so please think about members of your community when answering these questions.

22. Law and justice in your community Is there a serious problem in your community because of an inadequate system for law and justice, or because people do not know enough about their legal rights?	
23. Safety or protection from violence for women in your community Is there a serious problem for women in your community because of physical or sexual violence towards them, either in the community or in their homes?	
24. Alcohol or drug use in your community Is there a serious problem in your community because people drink a lot of alcohol, or use harmful drugs?	
25. Mental illness in your community Is there a serious problem in your community because people have a mental illness?	
26. Care for people in your community who are on their own Is there a serious problem in your community because there is not enough care for people who are on their own? For example, care for unaccompanied children, widows or elderly people, or unaccompanied people who have a physical or mental illness, or disability.	

Other serious problems:

Do you have any other serious problems that I have not yet asked you about? Write down the person's answers. 27.
28.
29.

Priority ratings for serious problems:

Read out the titles of all questions you have rated as '1', as well as any other serious problems listed above. Write down the person's answers (write down the number and title of the questions). 1.   Out of these problems, which one is the most serious problem?
2.   Which one is the second most serious problem?
3.   Which one is the third most serious problem?

* Throughout the HESPER form, the term 'community' should be replaced with the term that is most suitable to the local geographical area (for example village, town, neighbourhood, camp and so on).

Appendix 2

Humanitarian Emergency Settings Perceived Needs Scale (HESPER)

Training manual for interviewers
2011

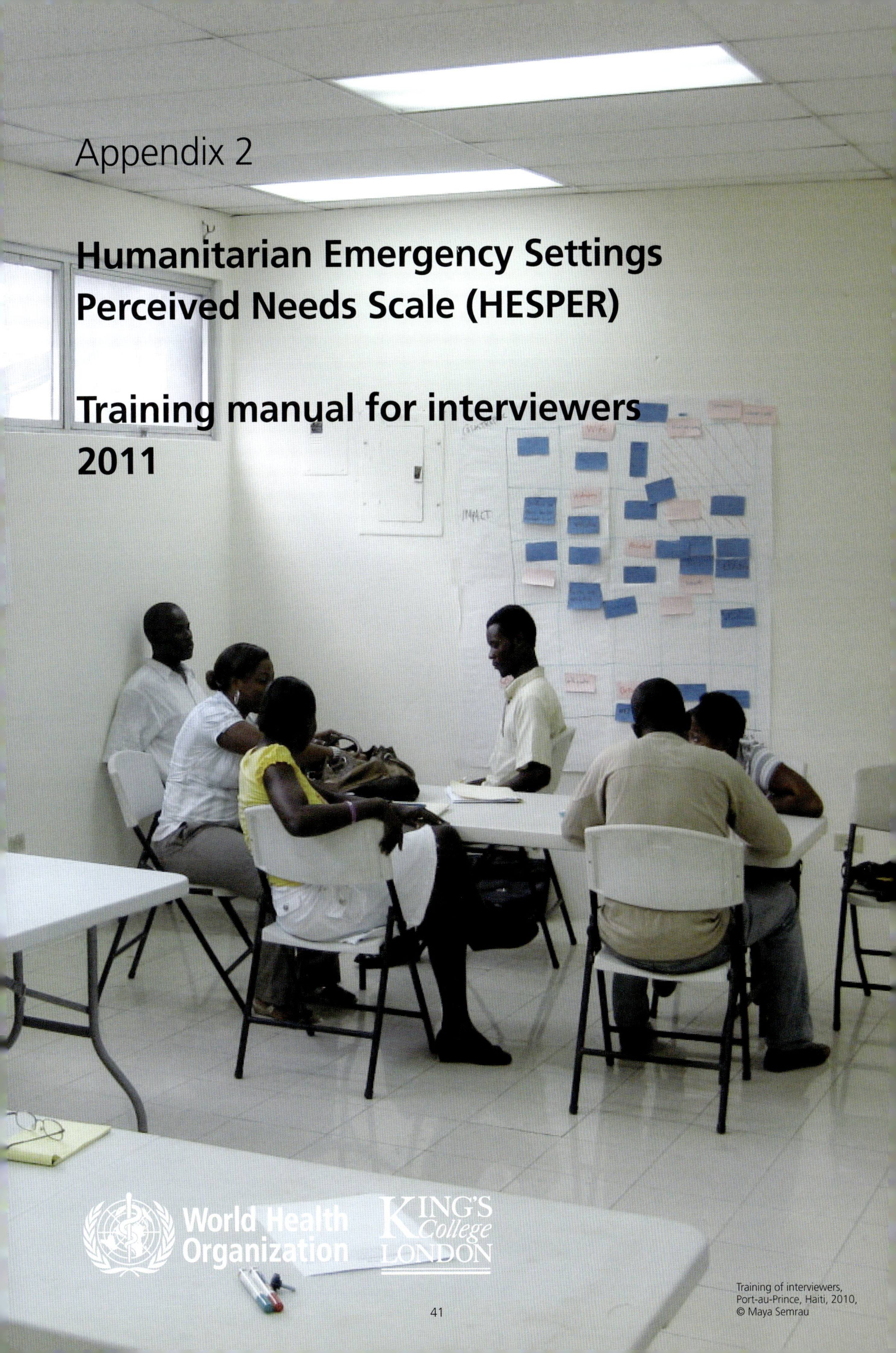

Training of interviewers,
Port-au-Prince, Haiti, 2010,
© Maya Semrau

Overview

This training manual explains how to use the HESPER Scale. It is written for interviewers or team leaders who would like to learn how to carry out a successful HESPER assessment. We recommend that you have this training manual with you during HESPER interviews.

Chapter 1 includes an introduction to the HESPER Scale and its rating system.

Chapter 2 provides an explanation of the whole HESPER assessment process. Chapters 2.1 and 2.2 contain the same information, but in different levels of detail – 2.1 gives a brief overview, and 2.2 provides a detailed explanation of the whole HESPER process. Section 2.2 is the most important section and you should read it thoroughly at least once before your first HESPER assessment. Section 2.3 provides further explanations for some of the HESPER questions.

Chapter 3 highlights various other things that are important to consider during a HESPER interview.

Chapter 4 provides examples to practice your HESPER interviewing skills.

Displacement camp in Port-au-Prince, Haiti, 2010, © Maya Semrau

Table of Contents

1 The HESPER Scale 44
   1.1 Introduction to the HESPER Scale 44
   1.2 Who is suitable as an interviewer? 44
   1.3 Rating the HESPER Scale 45

2 The HESPER interview 46
   2.1 Overview of the HESPER interview 46
   2.2 The HESPER assessment process in detail 46
   2.3 Explanations for individual HESPER questions 51

3 Other things to consider 55
   3.1 Safety 55
   3.2 Confidentiality 55
   3.3 Avoiding raised expectations 55
   3.4 Horrific events 55
   3.5 Handling distress 56
   3.6 Self-care 56
   3.7 Supervision 56

4 Practice assessments 57
   4.1 Example interview 57
   4.2 Practice questions 59
       Answers to practice questions 64
   4.3 Practice interviews 65
       Answers to practice interviews 68
   4.4 Practice role plays 69

1. The HESPER Scale

1.1 INTRODUCTION TO THE HESPER SCALE

There are many ways in which to assess people's needs. One way is to assess people's 'perceived needs'. These are needs which are felt or expressed by people themselves, and are problem areas they would like help with.

The Humanitarian Emergency Settings Perceived Needs Scale (HESPER) aims to provide a quick and reliable way of assessing the perceived serious needs of people affected by large-scale humanitarian emergencies such as war, conflict or major natural disaster.

The HESPER Scale assesses a wide range of social, psychological and physical problem areas. However, it does not provide an answer as to whether, or how to, offer help. It simply aims to identify those serious problems that are common in a population. These problems should then be assessed and addressed in more detail.

The HESPER Scale was developed by the World Health Organization in Geneva and King's College London to fill several gaps in the humanitarian field. We hope that it will enable needs assessments to be based directly on the views of people affected by humanitarian emergencies, and will help to provide a more accurate picture of the serious problems which the people affected by an emergency want help with. The opinions of many different people were collected whilst developing the HESPER Scale, including humanitarian experts, aid workers, refugees, and other local populations affected by humanitarian emergencies.

1.2 WHO IS SUITABLE AS AN INTERVIEWER?

It is important that interviewers are familiar with the local setting in which they are carrying out the assessment, and that the choice of interviewer is suitable to the local culture. For example, in some cultures it may be not be appropriate for a man to interview a woman, or for a younger woman to interview an older woman. Within the same country there may also be different cultural norms between particular groups, for example across different age groups, genders, or people of different religious beliefs. The choice of interviewer should be appropriate to the overall population and also to the particular group. If working in another culture, interviewers should make sure that their behaviour during the interview fits in with the cultural setting in which they are carrying out the interview. This includes, for example, dressing according to the cultural norms and acting in a way that is locally acceptable. Also, interviewers need to be good at communicating with other people and should have good basic interviewing skills, as well as some knowledge of ethical principles, such as understanding the importance of confidentiality and making sure that the person they are interviewing agrees to take part. We also recommend that interviewers have had an education of at least 12 years (that is, they have a high school diploma or equivalent).

As the interviewer, it is important that you are comfortable about doing the assessment. If you feel that you do not fit the criteria described above, or you feel uncomfortable about carrying out interviews after reading this manual, please let your project leader or supervisor know.

1.3 RATING THE HESPER SCALE

A HESPER assessment involves asking the person you are interviewing about 26 problem areas. You will rate whether the person feels that they have a serious problem in that particular area based on the answers they give. Before explaining the assessment process in more detail, it is important to understand how to make ratings on the HESPER Scale.

You should rate each question in the same way. You will ask the person about each problem area and give each question a rating based on their answers. See Box 1 for an explanation of the HESPER Scale's rating system.

Box 1

You should rate each question according to the following guidelines

Rate 9 (does not know / not applicable / declines to answer) if the person does not know how to answer the question, does not want to answer the question, or if the question does not apply to them.

Rate 1 (serious problem) if the person thinks that there is a serious problem for this question. A serious problem is a problem which the person feels is serious (however they define this).

Rate 0 (no serious problem) if the person does not think that there is a serious problem for this question.

2. The HESPER interview

2.1 OVERVIEW OF THE HESPER INTERVIEW

Each interview should take around 15 to 30 minutes, but this will vary.

There are six steps to each HESPER interview. These are outlined in Box 2 below.

Box 2

The six steps of a HESPER interview

1. **Before the interview:** Make sure you are familiar with the HESPER Scale and its rating system. You should have practiced this with your colleagues before your first HESPER assessment.

2. **Introduction to the interview:** Introduce yourself to the person you are interviewing, explain the purpose of the interview and the interview process, answer any questions they may have, and ask if they agree to take part. If they do agree to take part, make sure that they are comfortable and ready to start the interview (see Box 3 on page 47 for examples). Then write down the date, your name, the participant number, the location in which the person lives, as well as the person's gender and age at the top of the HESPER form.

3. **HESPER Scale – Need ratings:** Read out the text at the top of the HESPER form. Then ask questions about each of the HESPER Scale's problem areas and give each question a rating based on the person's answers (see Box 4 on page 48). Write the ratings in the appropriate column as you go along. Ask one or more follow-up questions for each area if necessary to make sure that you understand the person's views correctly.

4. **HESPER Scale – Other serious problems:** Once you have rated each of the HESPER Scale's problem areas, ask the person whether they have any other serious problems and write these down in the assigned spaces at the bottom of the HESPER form.

5. **HESPER Scale – Priority ratings for serious problems:** Then ask the person to tell you their three most serious problems in order of importance and write these down in the assigned spaces at the bottom of the HESPER form (see Box 6 on page 50 for an example).

6. **End of interview:** Thank the person for taking part in the interview, answer any questions they have, and make sure that they have your or your organisation's contact details (see Box 7 on page 51 for an example).

2.2 THE HESPER ASSESSMENT PROCESS IN DETAIL

The same six steps are now explained in more detail.

1. Before the interview

It is important that you are familiar with the HESPER Scale and its rating system before you start an assessment. Before your first assessment, take the time to practice carrying out interviews and making ratings. It may be a good idea to do this together with another interviewer using role plays (see sections 4.3 and 4.4 on pages 65 to 69). After you do a number of role plays, an experienced or knowledgeable supervisor should watch while you do another one. If you still feel unsure about anything afterwards, please ask your project leader or supervisor. It is a good idea for your supervisor to watch your first few 'real' interviews, if possible.

2. Introduction to the interview

During interviews, you should first introduce yourself to the person you are interviewing (give your name and say who you work for), explain the reasons for the interview, the interview process (including how long it will take), answer any questions the person may have, and ask whether they agree to take part. You should explain that participation is anonymous, completely voluntary, and that the person will receive no compensation or other benefits for taking part. Your supervisor or project leader may give you a participant information sheet to read out to the person you are interviewing that covers all these topics.

It is important that the person is comfortable and agrees to take part in the assessment. You should not pressurize the person into taking part. The person is free to choose whether or not to take part and they can end the interview at any point.

Throughout the assessment it is important to be friendly and respectful towards the person. This will help them to feel comfortable and give honest answers to your questions. You should also make sure that the person is comfortable with the place in which the assessment is being carried out. If possible, this should be somewhere private so that other people cannot overhear the interview. See Box 3 for examples of things you could say to make sure that the person feels comfortable before starting the interview.

Box 3

Examples of things to say to make sure the person feels comfortable

"Are you okay?" (or a cultural equivalent)

"Thank you for making time for me."

"Would you like some water?" (if the interview does not take place at the person's home)

"Are you ready to start the interview?"

Only go ahead with the interview (steps 3 to 6) if the person agrees to take part. You should write the date, your name, the location in which the person lives, as well as the person's gender and age at the top of the HESPER form. You should not write the person's name on the form but should use a pre-assigned participant number instead. You should have a separate sheet of paper which links the person's name with their participant number.

3. HESPER Scale – Need ratings

After reading out the text at the top of the HESPER form, rate each of the HESPER Scale's problem areas individually. You should go through the HESPER Scale and ask each question one after another. Write your ratings on the HESPER form for each question based on the person's answers. **It is important that you make a rating for each question on the HESPER form.**

You should assess each of the HESPER Scale's questions in the same way. You should make a rating for each question based on the person's answers according to the guidelines in Box 1 on page 45.

It is important that you read out the whole question to the person for each problem area. You do not need to read out the titles of the questions. If the person feels that there is a serious problem with any of the things mentioned in the question, you should rate it as '1' (serious problem).

Remember that you are rating the questions according to whether the person perceives there to be a _serious_ problem (however they define this).

It is also important that you rate each question according to what the person feels their serious problems are, and not what you think their serious problems are. You should record the person's views, even if you disagree with them. You should not let the person know if you disagree with them.

See Box 4 for an example of how to ask questions for each problem area and how to make ratings based on the person's answers. Please remember that this is a simplified example. Often, the person will not answer in such a clear-cut way. Although one question may be enough, you may sometimes have to ask more questions to be able to make a rating (see Chapter 4 for practice examples). You should try to keep your questions simple. It is also important that you understand the person's answers correctly. If you are not sure, ask them to explain their answer further. Listen carefully to what the person wants to say and make ratings based on this.

Box 4

Example of how to ask questions for each problem area

Interviewer: "Do you have a serious problem because you do not have a suitable place to live in?" (question 3)

Person being interviewed: "Yes."   (rate as 1 = serious problem)

"No."   (rate as 0 = no serious problem)

See Box 5 for tips on good interviewing techniques, which you should use during HESPER assessments.

Appendix 2
Training manual for interviewers

Box 5

Tips for good interviewing techniques

Be familiar with the HESPER Scale before you carry out an assessment. This will make you more confident and comfortable.

Interviewer's attitude

- Be warm and understanding towards the person you are interviewing. This means showing the person that you are listening, and responding to what they say in a kind and friendly way. You can say things like "That's great!", "I'm sorry!", "That's a shame!" or "I see!" to let the person know that you are listening and that you care about their situation. However, be careful not to make them believe that you will be able to help them with their situation.

- Let the person give you their opinion fully.

- Sometimes it may be useful to repeat the person's answers to them in your own words before making a rating, to make sure that you have understood them correctly. For example, you could say: "So, have I understood you correctly? Are you saying…?" This also shows the person that you are listening.

- It is possible that people you are interviewing may be angry or upset about their situation. Be understanding and kind.

Verbal skills

- Speak slowly and clearly to make sure that the person understands you.

- Use a pleasant and friendly voice.

- Leave many pauses and silences between questions. This gives the person the opportunity to think about their answers and to give you their opinion.

Phrasing questions

- Keep your questions simple and clear.

- If the person is not giving you a clear answer, try phrasing the question in a different way.

- It is okay to check with a person if their answer is not clear. For example, if you feel that the person is trying to say that they do not have a serious problem with 'Food' (question 2), but is not being clear about this, you could say: "So am I understanding correctly that you do not have a serious problem with food?"

- Sometimes using multiple choice questions can also be useful if the person is not giving a clear answer. For example, for the question 'The way aid is provided' (question 18), if the person has told you that some people are getting more aid than others, you could ask: "So is this a serious problem for you that some people are getting more aid than others, or is it not a serious problem?"

Handling diversions

- If the person gives very long or irrelevant answers, you can say something like: "That is very interesting. However, there are many more questions I have to ask, so would it be okay to please move on to those?" Or you could say: "We can talk about that some more after the interview, if you would like to."

- If the person starts asking for advice, information or your own personal experiences, you could say: "We are really interested to find out about your experiences and perceptions." Or you could say: "We can talk about that after the interview."

Handling distress *(also see section 3.5 on page 56)*

- If the person gets a little upset at any point during the assessment, slow down and take a short break if necessary. Ask the person whether they are okay to continue with the interview and stop the interview if they want to.

- Do not keep asking questions or challenge the person too much about sensitive or difficult subjects. If the person is getting very upset by a topic, it may be a good idea to close the interview booklet and be silent until they calm down. You could then say: "You seem very upset. Are you okay to continue with the interview, or would you prefer to stop?"

- Remember that the person can choose not to answer a question if they do not want to.

4. HESPER Scale – Other serious problems

After you have rated each problem area listed on the HESPER Scale, ask the person: "Do you have any other serious problems that I have not yet asked you about?" If the person has one or more other serious problems, write these down in the assigned spaces at the bottom of the form. You can ask the person to tell you up to three other serious problems.

5. HESPER Scale – Priority ratings for serious problems

You should then ask the person to tell you their three most serious problems (in order of priority).

Read out all the titles of the problem areas which the person has rated as 'serious problem' ('1' ratings), as well as any other serious problems listed under the 'Other serious problems' section. Then ask the person to list their three most serious problems in their order of importance, and write their answers in the assigned spaces. You should write down both the question number and title of the problem area. See Box 6 for an example of how to ask the person to prioritise their serious problems.

If the person has only told you that two problem areas are 'serious problems' ('1' ratings), ask them to rate those two problems in their order of importance. If the person has only told you about one (or no) serious problem, you do not need to fill in the 'Priority ratings' section as it is obvious that this problem is the most serious problem.

Box 6

Example of how to make priority ratings

Interviewer: "I am now going to read out all the problem areas which you have told me you have a serious problem with. I would like you to tell me which of these are your three most serious problems."

Read out all the areas which the person has told you they have a serious problem with (that is, all the questions you have rated as '1'), as well as any serious problems listed under the 'Other serious problems' section on the HESPER Scale (if this applies).

Then say: "Out of these problems, which one is the most serious problem?"

Person being interviewed: "Not having adequate health care." (question 9)

Interviewer: Write '9 - Health care' in the assigned space.

"Which one is the second most serious problem?"

Person being interviewed: "Not having a suitable place to live in." (question 3)

Interviewer: Write '3 - Place to live in' in the assigned space.

"Which one is the third most serious problem?"

Person being interviewed: Tells you a problem listed under the 'Other serious problems' section.

Interviewer: Write the problem listed under the 'Other serious problems' section in the assigned 'Priority ratings' space.

6. End of interview

At the end of the interview, thank the person for taking part, and ask whether they have any more questions or concerns. Take your time answering any questions before you leave and make sure that the person has got your organisation's or the project leader's contact details. See Box 7 for an example of how to end the interview.

> **Box 7**
>
> **Example of how to end a HESPER interview**
>
> "Thank you very much for taking part in this interview. I hope it was okay for you. I will now pass your answers on to (insert name of organisation) together with the answers of many other people from your community. We will not give your name to anybody and we will keep your answers safe and secure. As I mentioned before the interview, we are doing the assessment to find out what serious problems people in this community are facing. Have you got any more questions at this point?"
>
> "If you have any questions about this interview at any time in the future, please contact (insert the name of a person and organisation). Thank you again for taking part."

2.3 EXPLANATIONS FOR INDIVIDUAL HESPER QUESTIONS

In this section we explain some of the questions in the HESPER Scale. You may find this section useful if a person you are interviewing asks you to explain a question further, or if you are finding it difficult to decide on the correct rating for a question. You may use these explanations when a person asks you to explain a question – you should not offer your own interpretations. However, these explanations are not meant to be read out to every person you interview.

Some questions are about people's individual problems (for example, 'Drinking water' (question 1) and 'Food' (question 2)), while some are about the person's whole community (for example, 'Alcohol or drug use in your community' (question 24) and 'Mental illness in your community' (question 25)).

1. Drinking water

This question includes any water that is used either for drinking or for cooking. For example, this may include drinking water from taps in the person's home, water from shared taps, or bottled water. The question does not include water for washing (this is included under 'Keeping clean' (question 5)).

2. Food

This question is about whether the person has enough food, and also whether they have food that is appropriate and suitable to their needs. It also includes having suitable equipment and facilities to cook food, for example a stove, firewood, pots or pans.

3. Place to live in
This question may include a temporary or permanent house, hut, tent, or any other kind of shelter.

4. Toilets
This question refers to the toilet (or toilets) that the person uses on a regular basis. If they have a toilet in their home, this may refer to that toilet. If the person uses shared toilets (for instance in a camp setting), this may refer to the shared toilets.

5. Keeping clean
You should read out different questions for men and women. The question for women includes sanitary materials, whereas the question for men does not.

7. Income or livelihood
This question may include a wide range of problems to do with the person's livelihood, for example lack of income from employment, lack of access to farm land, lack of tools for farming, lack of boats or nets for fishing, or lack of access to other resources on which their livelihood depends.

8. Physical health
This question may include any kind of physical illnesses or injuries, including physical disabilities.

9. Health care
You should read out different questions for men and women. Any kind of health care is included, for example hospital treatment, access to a doctor or nurse, access to medications, and sexual health care (including access to contraceptives). The question for women also includes access to support and health care during pregnancy and childbirth.

11. Safety
This question may refer to any serious problem that the person has with safety or security. This could include if they do not feel safe because of crime, conflict, war, violence, the political situation, or any other kind of instability. The question asks about the person's family, so it includes their children, husband or wife, or other family members.

12. Education for your children
If the person has already told you at the beginning of the interview that they do not have any children, you do not need to ask this question and can give a rating of 'not applicable' (9). If the person only has children who are not of school age (that is, they are either too old or too young to go to school), you should also rate the question as 'not applicable' (9).

Appendix 2
Training manual for interviewers

14. Support from others
This question refers both to emotional and practical support. Practical support may include financial help, help with daily living, help with transport, help with babysitting, or any other kind of practical help. Emotional support is any support offered by another person that helps the person you are interviewing deal with any difficult emotions they may experience. For example, this may include someone talking to the person about their problems, or someone showing that they care about the person's difficulties.

15. Separation from family members
This may include, for example, if the person is separated from their family members because they (or their family) have been forced to leave their home, the person does not know where one or more of their family members are, a family member is missing, or the person is not able to leave the place where they are living to visit family members.

16. Being displaced from home
This question refers to any serious problem that the person is having because they have been displaced from their home country, or home city or village. If the person has already told you that they have not been displaced from home, you do not need to ask this question, and can rate the question as 'not applicable' (9). You should rate the question as 'serious problem' (1) if the person feels that they have a serious problem because they have had to leave their home environment. You should rate it as 'no serious problem' (0) if the person feels that they do not have a serious problem because of this. Please remember that you are not rating whether the person has been displaced from home, but rather whether they feel they have a serious problem because they have had to leave their home. It is therefore possible that the person has had to leave their home, but does not feel that this is a serious problem (you would then give a rating of '0').

17. Information
You should read out different questions for people who have been displaced from home and people who have not been displaced from home.

18. The way aid is provided
It is important to remember that you are assessing whether the person thinks that there is a serious problem in this area. You should rate the question as 'no serious problem' (0) if the person feels that aid is being, or has been, handed out fairly, and that aid agencies are involving the community in the aid process. You should also rate this question as 'no serious problem' (0) if the person does not have fair access to aid, or the community is not involved in the aid process, but the person does not see this as a serious problem.

19. Respect

This question includes any disrespect or humiliation felt by the person, for example because of aid workers, people in the person's community or family, or the situation in which the person lives.

20. Moving between places

This may include serious problems with moving between places because of problems with transport, because the person thinks that moving between places is unsafe, or because they have physical problems that stop them from moving around.

24. Alcohol or drug use in your community

This question may include harmful drugs that can be bought from pharmacies or other shops, as well as illegal drugs.

25. Mental illness in your community

This question may refer to any mental illnesses or mental health problems that people in the community are experiencing. It is important to remember that you are not assessing whether these mental illnesses exist in the community, but rather whether the person feels there is a serious problem in the community because people have a mental illness (however the person defines this). For example, if the person thinks that many people in the community have a mental illness, but does not think that this is a serious problem, you should rate the question as 'no serious problem' (0). However, if the person thinks that people in the community have a mental illness, and thinks that this is a serious problem, you should rate the question as 'serious problem' (1).

Displaced persons camp in Mogadishu, Somalia, 2000/2001, © WHO

3. Other things to consider

3.1 SAFETY

It is important that both you and the person you are interviewing are safe throughout the interview and feel comfortable about the place in which the interview is being held. You should choose a setting which is safe and culturally appropriate. For example, it may sometimes not be suitable to do the interview in the person's house or shelter. In this case arrangements should be made for the interview to take place in a quiet and suitable place. Always make sure that somebody knows where and when you are doing an interview. If possible, carry a mobile phone or satellite phone with you. Depending on the situation, it may sometimes be necessary or advisable to do the interviews in pairs or to have someone else with you.

3.2 CONFIDENTIALITY

In order to respect the person's right to privacy, it is important that you keep their details and answers confidential. This means that you should not show or discuss their answers or personal details with other people outside the assessment team. You should not discuss anything with others, even after the assessment has been carried out. You should do the interview in a place which is as private as possible. Ideally this means that the interview should be in a quiet room with only the interviewer and the person being interviewed present. However, this may not always be possible or culturally appropriate.

3.3 AVOIDING RAISED EXPECTATIONS

Sometimes when people take part in interviews, they mistakenly assume that the assessment team will be able to help them with their problems. It is important that you make sure people understand that they will get no direct benefits (for them or their family) by taking part in the interview. You should make it clear throughout the interview that they will not receive any compensation, extra aid, or other benefits just by talking to you. This is important so that people's expectations for help are not raised, and also so that they do not pretend that their needs are more serious than they actually are.

3.4 HORRIFIC EVENTS

Thinking about violent or other horrific events can cause people to become distressed. You should not ask about these events in detail. The HESPER Scale is specifically designed not to need a great level of detail. If the person you are interviewing wants to talk about these events, allow them to do so to some extent, but do not ask them for more details as this is not the purpose of doing the HESPER assessment. In any case, be patient and show that you are listening.

3.5 HANDLING DISTRESS

The person you are interviewing may stop the interview at any time. If they ask to stop the interview, please do so immediately. The person does not need to give a reason for wanting to stop the interview. It is okay to continue with the interview if the person is a little upset and agrees to gently continue with the interview. However, if the person is getting very upset by a topic, it may be a good idea to close the interview booklet and be silent until they calm down. You could then say: "You seem very upset. Are you okay to continue the interview or would you prefer to stop?" At the end of the interview, refer the person to the best available psychosocial support worker and let your project leader or supervisor know. Before your first interview your supervisor should give you a list of support organisations that you can give to the people you interview. See Box 8 for an example of how you could end the interview, if the person chooses to end the interview early.

Box 8

Example of how to end the interview early, if requested by the person being interviewed

Interviewer: "You seem very upset. Are you okay to continue the interview or would you prefer to stop?"

Person being interviewed: "I would prefer to stop."

Interviewer: "Okay, that is no problem. We will stop the interview. Thank you very much for taking part in the assessment. I am very sorry that you got upset. If you want, I can let somebody know that you are very upset by the situation you are in and they may contact you to talk about this. Would that be okay with you?

I will now pass your answers to (insert name of organisation) together with the answers of many other people from your community. We will not give your name to anybody and we will keep your answers safe and secure. As I mentioned before the interview, we are doing the assessment to find out what serious problems people in this community are facing. Have you got any more questions at this point?

If you have any questions about this interview at any time in the future, please contact (insert the name of a person and organisation). Thank you again for taking part."

3.6 SELF-CARE

It is possible that you may feel upset or distressed by an interview, or that you find the interview process difficult. If this is the case, please speak to your supervisor or project manager, or a staff welfare officer if there is one available.

3.7 SUPERVISION

If possible, you should meet with your supervisor and other interviewers at the end of each day to review the interview process.

Appendix 2
Training manual for interviewers

4. Practice assessments

4.1 EXAMPLE INTERVIEW

It is important to practice interviews before using the HESPER Scale for the first time. Here are some examples of the types of questions and answers which may come up during an interview. Only a few of the HESPER Scale's problem areas are given here as examples.

Mani is a 42-year-old man from the Democratic Republic of the Congo (DRC). He and his wife, five children and other family members have had to leave their village due to rebel fighting in the area.

Drinking water (question 1)
Interviewer: "Do you have a serious problem because you do not have enough water that is safe for drinking or cooking?"
Mani: "We had problems for a long time and we had to find water wherever we could. In the last few days though aid workers have come and they have given us water to drink."
Interviewer: "That's good. So would you say that you have still got a serious problem with this, or is it okay now?"
Mani: "It is a problem, but it is not a serious problem."

Food (question 2)
Interviewer: "Do you have a serious problem with food? For example, because you do not have enough food, or good enough food, or because you are not able to cook food."
Mani: "We don't have enough food at all."
Interviewer: "Would you say that this is a serious problem?"
Mani: "Yes, very serious."

Place to live in (question 3)
Interviewer: "Do you have a serious problem because you do not have a suitable place to live in?"
Mani: "We have nowhere to stay at the moment. We are sleeping outside without any shelter. My children are cold and when it rains we get wet. It is a serious problem."

Clothes, shoes, bedding or blankets (question 6)
Interviewer: "Do you have a serious problem because you do not have enough, or good enough, clothes, shoes, bedding or blankets?"

Mani: "We have received some from the aid organisation."
Interviewer: "That's great. So, would you say that you are okay with it now, or do you still think that you have a serious problem with it?"
Mani: "No, we are okay with it."

Separation from family members (question 15)
Interviewer: "Do you have a serious problem because you are separated from family members?"
Mani: "My family have been lucky in that way. We have all managed to stay together the whole time."
Interviewer: "That is great. So, would you say that you have a serious problem in that area, or is it okay?"
Mani: "No, that is okay."

This is how you would rate each question based on Mani's answers.

Rating: 0 = no serious problem 1 = serious problem 9 = does not know / not applicable / declines to answer	Ratings
1. **Drinking water** Do you have a serious problem because you do not have enough water that is safe for drinking or cooking?	0
2. **Food** Do you have a serious problem with food? For example, because you do not have enough food, or good enough food, or because you are not able to cook food.	1
3. **Place to live in** Do you have a serious problem because you do not have a suitable place to live in?	1
6. **Clothes, shoes, bedding or blankets** Do you have a serious problem because you do not have enough, or good enough, clothes, shoes, bedding or blankets?	0
15. **Separation from family members** Do you have a serious problem because you are separated from family members?	0

After the interviewer has rated all of the HESPER Scale's problem areas based on Mani's answers, she asks him the following questions.

Interviewer: "Do you have any other serious problems that I have not yet asked you about?"
Mani: "No, I think we have talked about all of my problems."
Interviewer: "So, you have told me that you have a serious problem with 'Food' and 'Place to live in'. Out of these problems, which one is the most serious problem?"
Mani: "Our most serious problem is not having enough food. Our next biggest problem is not having a suitable place to live in."

Appendix 2
Training manual for interviewers

This is how you would make priority ratings based on Mani's answers.

Priority ratings for serious problems:
Read out the titles of all questions you have rated as '1', as well as any other serious problems listed above. Write down the person's answers (write down the number and title of the questions).
1. Out of these problems, which one is the most serious problem?   *2 – Food*
2. Which one is the second most serious problem?   *3 – Place to live in*

4.2 PRACTICE QUESTIONS

This section gives you the chance to practice your interviewing skills. Each question has a set of multiple choice answers. Please try to answer the questions before looking at the answers on page 64.

---

1. Interviewer: "Do you have a serious problem because you do not have enough, or good enough, clothes, shoes, bedding or blankets?" (question 6)
   Person being interviewed: "Yes."

   What should the interviewer do?
   a. Rate the question as '1' and move on to the next question.
   b. Rate the question as '0' and move on to the next question.
   c. Rate the question as '9' and move on to the next question.

---

2. Interviewer: "Do you have a serious problem because you do not have a suitable place to live in?" (question 3)
   Person being interviewed: "No, we have a hut. That is okay."

   What should the interviewer do?
   a. Rate the question as '0' and move on to the next question.
   b. Rate the question as '9' and move on to the next question.
   c. Rate the question as '1' and move on to the next question.

The Humanitarian Emergency Settings Perceived Needs Scale (HESPER): Manual with Scale

3. Interviewer: "Is there a serious problem in your community because people have a mental illness?" (question 25)
   Person being interviewed: "I am not sure. There may be, but I don't know."

   What should the interviewer do?
   a. Rate the question as '0' and move on to the next question.
   b. Rate the question as '1' and move on to the next question.
   c. Rate the question as '9' and move on to the next question.

4. Interviewer: "Is there a serious problem in your community because there is not enough care for people who are on their own? For example, care for unaccompanied children, widows or elderly people, or unaccompanied people who have a physical or mental illness, or disability." (question 26)
   Person being interviewed: "There are many people who are not looked after. It is a serious problem."

   What should the interviewer do?
   a. Rate the question as '1' and move on to the next question.
   b. Rate the question as '9' and move on to the next question.
   c. Rate the question as '0' and move on to the next question.

5. Interviewer: "Do you have a serious problem with food? For example, because you do not have enough food, or good enough food, or because you are not able to cook food." (question 2)
   Person being interviewed: "No, it is not a problem."

   What should the interviewer do?
   a. Rate the question as '1' and move on to the next question.
   b. Rate the question as '0' and move on to the next question.
   c. Rate the question as '9' and move on to the next question.

6. Interviewer: "Do you have a serious problem because you do not have easy and safe access to a clean toilet?" (question 4)
   Person being interviewed: "No, that is okay."

   What should the interviewer do?
   a. Rate the question as '9' and move on to the next question.
   b. Rate the question as '0' and move on to the next question.
   c. Rate the question as '1' and move on to the next question.

Appendix 2
Training manual for interviewers

7. Interviewer: "Do you have a serious problem because your children are not in school, or are not getting a good enough education?" (question 12)
   Person being interviewed: "I do not have any children."

   What should the interviewer do?
   a. Rate the question as '0' and move on to the next question.
   b. Rate the question as '1' and move on to the next question.
   c. Rate the question as '9' and move on to the next question.

8. Interviewer: "Do you have a serious problem because in your situation it is difficult to care for family members who live with you? For example, young children in your family, or family members who are elderly, physically or mentally ill, or disabled." (question 13)
   Person being interviewed: "That is family business. I don't want to talk about this."

   What should the interviewer do?
   a. Rate the question as '1' and move on to the next question.
   b. Rate the question as '0' and move on to the next question.
   c. Rate the question as '9' and move on to the next question.
   d. Say: "This seems to be an issue. It would be great if you could please give me an answer to this question. Do you have a serious problem with this?"

9. Interviewer: "Do you have a serious problem because you do not have enough information? For example, because you do not have enough information about the aid that is available." (question 17)
   Person being interviewed (non-displaced): "No, I would say that is okay."

   What should the interviewer do?
   a. Rate the question as '0' and move on to the next question.
   b. Rate the question as '1' and move on to the next question.
   c. Rate the question as '9' and move on to the next question.

10. Interviewer: "Do you have a serious problem because you are not able to move between places? For example, going to another village or town." (question 20)
Person being interviewed: "It is a problem. Sometimes it can be difficult to get to my workplace because of road blocks."

What should the interviewer do?
   a. Rate the question as '1' and move on to the next question.
   b. Ask: "Would you say that this is a **serious** problem?" If the person answers "Yes", rate the question as '1' and move on to the next question. If they answer "No", rate the question as '0' and move on to the next question.
   c. Rate the question as '0' and move on to the next question.

11. Interviewer: "Do you have a serious problem because you do not have enough water that is safe for drinking or cooking?" (question 1)
Person being interviewed: "Yes, it is – we have to collect water from the rain and it does not rain often. The water is very dirty. Sometimes we do not have water for a few days."

What should the interviewer do?
   a. Rate the question as '1' and move on to the next question.
   b. Rate the question as '0' and move on to the next question.
   c. Rate the question as '9' and move on to the next question.
   d. Ask: "Is this a **serious** problem?" If the person answers "Yes", rate the question as '1' and move on to the next question. If they answer "No", rate the question as '0' and move on to the next question.

12. Interviewer: "Do you have a serious problem because you or your family are not safe or protected where you live now? For example, because of conflict, violence or crime in your community, city or village." (question 11)
Person being interviewed: Looks upset and cries a little.

What should the interviewer do?
   a. Ask: "Is the problem serious?" If the person answers "Yes", rate the question as '1' and move on to the next question. If they answer "No", rate the question as '0' and move on to the next question.
   b. Ask: "Are you okay to continue?" If the person says "Yes", ask: "Is the problem serious?" If they answer "Yes", rate the question as '1' and move on to the next question. If they answer "No", rate the question as '0' and move on to the next question.
   c. Rate the question as '1' and move on to the next question.

Appendix 2
Training manual for interviewers

13. Interviewer: "Do you have a serious problem because you have too much free time in the day?" (question 21)
    Person being interviewed: "It is a problem – there is not much to do."

    What should the interviewer do?
    a. Rate the question as '1' and move on to the next question.
    b. Ask: "Would you say that it is a **serious** problem?" If the person answers "Yes", rate the question as '1' and move on to the next question. If they answer "No", rate the question as '0' and move on to the next question.
    c. Rate the question as '0' and move on to the next question.
    d. Rate the question as '9' and move on to the next question.

14. Interviewer: "Do you have a serious problem because of inadequate aid? For example, because you do not have fair access to the aid that is available, or because aid agencies are working on their own without involvement from people in your community." (question 18)
    Person being interviewed: "I have not heard anything. Maybe you can tell me. What is happening? Are we going to receive any aid? And where do we get it from? We haven't been told anything."

    What should the interviewer do?
    a. Say: "That's a shame. Unfortunately I do not have much information either. I am happy to talk about this after the interview. We would really like to hear about your experiences at this time. Would it be okay to continue with the interview for now? So, would you consider the lack of information a serious problem?"
    b. Ask: "Is this lack of information a serious problem?"
    c. Rate the question as '1' and move on to the next question.
    d. Rate the question as '0' and move on to the next question.

15. Interviewer: "Is there a serious problem for women in your community because of physical or sexual violence towards them, either in the community or in their homes?" (question 23)
    Person being interviewed (female): Gets upset and sobs heavily.

    What should the interviewer do?
    a. Ask: "Is this a serious problem?"
    b. Close the interview booklet. Wait until the sobbing stops. Ask: "Are you okay? Are you okay to continue with the interview?" If the person answers "Yes", open the interview booklet and continue the interview. If they answer "No", end the interview.
    c. Ask: "Are you okay? Are you okay to continue with the interview?" If the person answers "Yes", continue with the interview. If they answer "No", end the interview.
    d. Rate the question as '1' and move on to the next question.

Answers to practice questions

1. The correct answer is a. If the person thinks there is a serious problem, the interviewer should rate the question as '1'.

2. The correct answer is a. If the person does not think there is a serious problem, the interviewer should rate the question as '0'.

3. The correct answer is c. If the person does not know how to answer, the interviewer should rate the question as '9'.

4. The correct answer is a. If the person thinks there is a serious problem, the interviewer should rate the question as '1'.

5. The correct answer is b. If the person does not think there is a serious problem, the interviewer should rate the question as '0'.

6. The correct answer is b. If the person does not think there is a serious problem, the interviewer should rate the question as '0'.

7. The correct answer is c. If a question does not apply to the person, the interviewer should rate the question as '9'.

8. The correct answer is c. The person can choose not to answer a question. If the person does not want to answer a question, the interviewer should rate that question as '9'.

9. The correct answer is a. If the person does not think there is a serious problem, the interviewer should rate the question as '0'.

10. The correct answer is b. The interviewer should only rate the question as '1' if the person thinks the problem is serious.

11. The correct answer is a. If the person thinks there is a serious problem, the interviewer should rate the question as '1'.

12. The correct answer is b. If the person gets upset, the interviewer should make sure that they are okay to continue with the interview.

13. The correct answer is b. The interviewer should only rate the question as '1' if the person thinks the problem is serious.

14. The correct answer is a. If the person answers "Yes", the interviewer should rate the question as '1'. If the person answers "No", the interviewer should rate the question as '0'.

15. The correct answer is b. It is probably best to close the interview booklet, let the person cry, and get permission to restart the interview when they stop crying. The person can stop the interview at any time without having to give a reason.

4.3 PRACTICE INTERVIEWS

In this section you can find examples of answers that people may give during HESPER assessments (for half of the HESPER questions). You may use these examples when practicing interviews. You can find the answers on pages 68 to 69.

Individual questions

1. Drinking water

Interviewer: "Do you have a serious problem because you do not have enough water that is safe for drinking or cooking?"
Person being interviewed: "Yes"
What should the interviewer do next?

2. Food

Interviewer: "Do you have a serious problem with food? For example, because you do not have enough food, or good enough food, or because you are not able to cook food."
Person being interviewed: "No."
What should the interviewer do next?

3. Place to live in

Interviewer: "Do you have a serious problem because you do not have a suitable place to live in?"
Person being interviewed: "I have a house. It is not great but it is okay."
What should the interviewer do next?

4. Toilets

Interviewer: "Do you have a serious problem because you do not have easy and safe access to a clean toilet?"
Person being interviewed: "I can use the toilets in the camp, but I am scared of going there. It is very dark at night."
What should the interviewer do next?

The Humanitarian Emergency Settings Perceived Needs Scale (HESPER): Manual with Scale

5. Keeping clean

Interviewer: "Do you have a serious problem because in your situation it is difficult to keep clean? For example, because you do not have enough soap, water or a suitable place to wash."
Person being interviewed (male): "I don't want to talk about that. That is private."
What should the interviewer do next?

6. Clothes, shoes, bedding or blankets

Interviewer: "Do you have a serious problem because you do not have enough, or good enough, clothes, shoes, bedding or blankets?"
Person being interviewed: "Could you tell me where I can find some? My clothes are very old."
What should the interviewer do next?

7. Income or livelihood

Interviewer: "Do you have a serious problem because you do not have enough income, money or resources to live?"
Person being interviewed: "Yes."
What should the interviewer do next?

8. Physical health

Interviewer: "Do you have a serious problem with your physical health? For example, because you have a physical illness, injury or disability."
Person being interviewed: "My leg hurts."
What should the interviewer do next?

9. Health care

Interviewer: "Do you have a serious problem because you are not able to get adequate health care for yourself? For example, treatment or medicines, or health care during pregnancy or childbirth."
Person being interviewed (female): Looks upset and cries a little.
What should the interviewer do next?

10. Distress

Interviewer: "Do you have a serious problem because you feel very distressed? For example, very upset, sad, worried, scared, or angry."
Person being interviewed: "No, I am okay."
What should the interviewer do next?

11. Safety

Interviewer: "Do you have a serious problem because you or your family are not safe or protected where you live now? For example, because of conflict, violence or crime in your community, city or village."
Person being interviewed: Gets upset and cries heavily.
What should the interviewer do next?

12. Education for your children

Interviewer: "Do you have a serious problem because your children are not in school, or are not getting a good enough education?"
Person being interviewed: "I don't have any children."
What should the interviewer do next?

Community questions

22. Law and justice in your community

Interviewer: "Is there a serious problem in your community because of an inadequate system for law and justice, or because people do not know enough about their legal rights?"
Person being interviewed: "I don't think so, no."
What should the interviewer do next?

23. Safety or protection from violence for women in your community

Interviewer: "Is there a serious problem for women in your community because of physical or sexual violence towards them, either in the community or in their homes?"
Person being interviewed: "I don't know about that."
What should the interviewer do next?

24. Alcohol or drug use in your community

Interviewer: "Is there a serious problem in your community because people drink a lot of alcohol, or use harmful drugs?"
Person being interviewed: "I have heard some people saying that it is a problem, but I don't think it is a problem."
What should the interviewer do next?

Answers to practice interviews

Individual questions

1. The interviewer should rate the question as '1' ('serious problem').

2. The interviewer should rate the question as '0' ('no serious problem').

3. The interviewer should say: "Would you say this is a **serious** problem?" If the person answers "Yes", the interviewer should rate the question as '1' ('serious problem'). If the person answers "No", the interviewer should rate the question as '0' ('no serious problem').

4. The interviewer should say: "Is this a serious problem for you?" If the person answers "Yes", the interviewer should rate the question as '1' ('serious problem'). If the person answers "No", the interviewer should rate the question as '0' ('no serious problem').

5. The interviewer should rate the question as '9' ('does not know / not applicable / declines to answer').

6. The interviewer should say something like: "I am sorry to hear that. Unfortunately I do not have any information about that, but I am happy to talk about this after the interview. Would it be okay to continue with the interview for now? Do you think it is a serious problem that your clothes are old?" If the person answers "Yes", the interviewer should rate the question as '1' ('serious problem'). If the person answers "No", the interviewer should rate the question as '0' ('no serious problem').

7. The interviewer should rate the question as '1' ('serious problem').

8. The interviewer should ask: "Is this a serious problem?" If the person answers "Yes", the interviewer should rate the question as '1' ('serious problem'). If the person answers "No", the interviewer should rate the question as '0' ('no serious problem').

9. The interviewer should ask: "Are you okay to continue?" If the person says "Yes", the interviewer should ask: "Is this a serious problem?" If the person answers "Yes", the interviewer should rate the question as '1' ('serious problem'). If the person answers "No", the interviewer should rate the question as '0' ('no serious problem').

10. The interviewer should rate the question as '0' ('no serious problem').

11. The interviewer should close the interview booklet and wait until the sobbing stops, then ask: "Are you okay? Are you okay to continue with the interview?" If the person says "Yes", the interviewer should open the interview booklet and continue the interview. If the person says "No", the interviewer should end the interview.

12. The interviewer should rate the question as '9' ('does not know / not applicable / declines to answer').

Community questions

22. The interviewer should rate the question as '0' ('no serious problem').

23. The interviewer should rate the question as '9' ('does not know / not applicable / declines to answer').

24. The interviewer should rate the question as '0' ('no serious problem'), as you are rating whether the person you are interviewing thinks this is a serious problem, not whether anybody else thinks this is a serious problem.

4.4 PRACTICE ROLE PLAYS

You should now practice HESPER assessments through role play.

It is a good idea to practice at least three HESPER assessments with a colleague before you do an actual interview. To practice, one of you should act as the interviewer and the other as the person being interviewed – then swap roles. A third colleague may watch the role play and give feedback. If you are in the role of the person being interviewed, you should try to give easy answers at first if the interviewer has never done this type of interview before. Then over time, you can start to challenge the interviewer by giving more difficult answers.

Thatta, Sindh, Pakistan, 2008,
© Sandie Walton-Ellery

Appendices

Appendix 3

Example HESPER Report
(written in plain language)

The Perceived Needs of the Population in *Location X*
Month 20XX

Name of author
Affiliation
email: *author@institution.org*

Coordination of implementation of study:
Name agency (name relevant agency staff)

Technical support: *Name agency (name relevant agency staff)*

Financial support: *Name source*

With special thanks to the interviewed people in *Location X*.

The views expressed in this report do not necessarily represent the decisions, policies, or views of the agencies that are associated with this assessment.

INTRODUCTION
Describe the context in one paragraph (for example nature and size of the humanitarian emergency and its response).

This report aims to present the findings of an assessment of the perceived needs of the population in *Location X* to relevant stakeholders.

OVERVIEW OF STUDY
The goals of this study were as follows:
1. The first goal was to find out the perceived needs (i.e. the serious problems) that adults living in *Location X* have.

   To measure their serious problems the Humanitarian Emergency Settings Perceived Needs Scale (HESPER) was used. The HESPER Scale measures the serious problems of adults living in humanitarian situations (for instance during conflicts or other disasters), based directly on their own views (i.e. people's perceived needs). It shows the problem areas with which people would like help. The HESPER Scale aims to provide a quick, scientifically robust way to measure people's serious problems, and includes a wide range of social, psychological and physical problem areas.

2. The second goal was to compare the perceived needs that different groups of people have, for example men versus women.

SAMPLING METHOD
Describe sampling methods in one paragraph. Mention what percentage of people who were invited to participate agreed to take part.

SAMPLE
In total, 269 participants were interviewed. All participants in the study were over 18 years of age (the oldest participant was 84 years old). Table 1 shows the characteristics of study participants.

PROCEDURE
Six local interviewers interviewed the participants in *name language* after receiving training on the use of the HESPER Scale. Interviews took place between *date* and *date*, and took place in participants' own homes.

**Appendix 3
Example HESPER Report**

Table 1: Demographic characteristics of study participants. Figures are displayed as number of participants (% in brackets), or averages (means).

	Total (n=269)
Sex	
Men	139 (51.7%)
Women	130 (48.3%)
Average age	36.9
Marital status	
Married	217 (80.7%)
Unmarried	50 (18.6%)
Divorced	2 (0.7%)
Average number of children	2.4
Level of education	
Illiterate / No formal education	108 (40.1%)
Primary school (grades 1 to 5)	55 (20.4%)
Secondary school (grades 6 to 10)	61 (22.7%)
Intermediate (grades 11 to 12)	36 (13.4%)
University	9 (3.3%)
Employment status	
Employed	135 (50.2%)
Not employed	134 (49.8%)
Religion	
Religion 1	178 (66.2%)
Religion 2	52 (19.3%)
Religion 3	18 (6.7%)
Religion 4	16 (5.9%)
Other religion*	5 (1.9%)
Average years displaced	19.0

* Other religions include *insert other religions*.

KEY FINDINGS

1. Interviewers used the HESPER Scale to ask participants about 26 different types of problems (i.e. problem areas). Overall, participants rated 8.1 of these areas as serious problem (the lowest number was 0 and the highest was 21). Figure 1 shows an overview of the number of areas rated as serious problem by participants.

Table 2 and Figure 2 show the frequency with which each of the 26 HESPER areas were rated as one of participants' three most serious problems (i.e. as either their most serious problem, second most serious problem, or third most serious problem). 'Income or livelihood' was rated by almost half of all participants (47.2%) as one of their three most serious problems, more than any other problem area. Other areas which were named by more than 10% of participants as one of their three most serious problems included 'Food' (24.5%), 'Physical health' (23.0%), 'Place to live in' (20.8%), 'Being displaced from home' (18.6%), 'Separation from family members' (16.7%), 'Clothes, shoes, bedding or blankets' (16.4%), and 'Alcohol or drug use in your community' (14.5%).

Table 3 shows the number of participants who rated each of the 26 HESPER areas as serious problem. 'Income or livelihood' was rated as serious problem by 75.1% of participants, again more than any other problem area. The following areas were rated as serious problem by around half of participants: 'Food' (58.0%), 'Being displaced from home' (52.0%), 'Clothes, shoes, bedding, or blankets' (49.1%), 'Place to live in' (44.6%), and 'Separation from family members' (42.0%).

When asked to name any other serious problems not listed on the HESPER Scale, 44 (16.4%) participants named a problem related to resettlement, of whom 13 (4.8%) rated it as one of their three most serious problems.

2. Men and women had a similar number of serious problems overall; the average number for men was 8.5 and for women was 7.7. This difference was not statistically significant.

CONCLUSIONS

1. The study gives an overview of the serious problems that the population living in *Location X* have, based directly on their own views. 'Income or livelihood' was the area which was perceived as serious problem by the largest number of participants, and was also rated by the largest number of participants as one of their three most serious problems. Other areas which were commonly rated as one of participants' three most serious problems, and were also perceived as serious problem by a large number of participants, were 'Food', 'Physical health', 'Place to live in', 'Being displaced from home', 'Separation from family members', 'Clothes, shoes, bedding, or blankets', and 'Alcohol or drug use in your community'.

Other serious problems which were named commonly by participants were issues with resettlement.

2. Men and women had a similar number of serious problems overall.

Appendix 3
Example HESPER Report

LIMITATIONS

No substantial errors or biases were indentified by the research team. Indeed, the sample size was large, sampling was representative, interviewers were well trained, reliability and validity data were good, and interviewers did not report that participants had an inclination to overestimate or underestimate their needs.

RECOMMENDATIONS

Based on our findings, we recommend that:
1. Actors in *Location X* should consider addressing *name prioritized perceived need areas*.
2. More detailed interviews (for example key informant interviews or focus groups) should be conducted with the population in *Location X*. These should focus especially on *insert perceived need areas* to gain a deeper understanding of them, and to identify relevant community resources, suitable interventions and supports.

Figure 1: Number of serious problem ratings by number of participants.

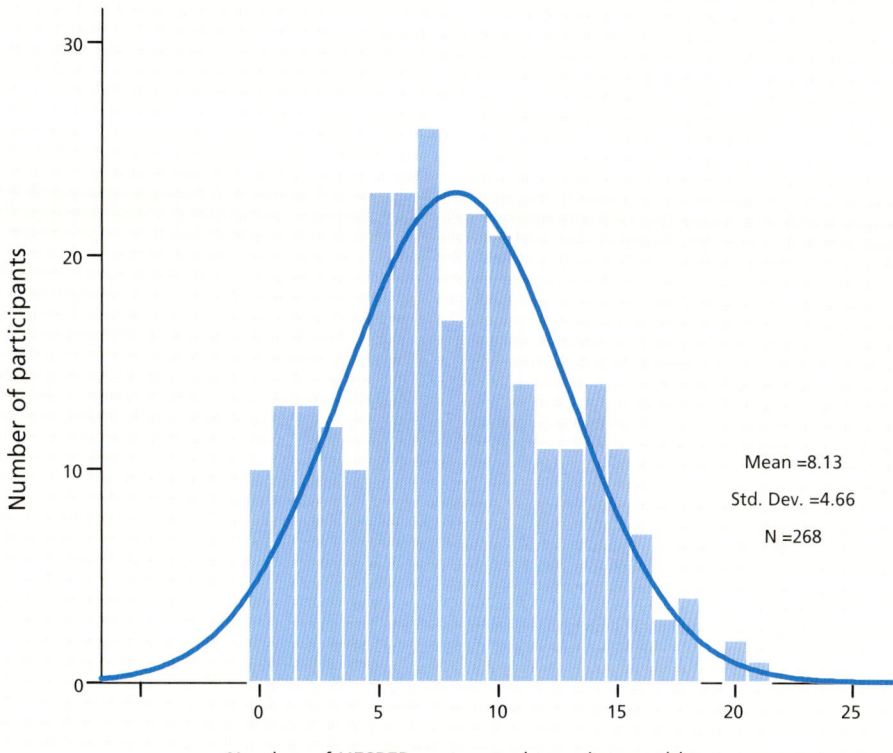

Table 2: Number of participants (% in brackets) who rated each of the HESPER Scale's problem areas as one of their three most serious problems (n=269). Items are ranked and listed in descending order of total priority ratings.

HESPER item	Total priority ratings	Priority rating 1	Priority rating 2	Priority rating 3
1. Income or livelihood	127 (47.2%)	57 (21.2%)	38 (14.1%)	32 (11.9%)
2. Food	66 (24.5%)	28 (10.4%)	20 (7.4%)	18 (6.7%)
3. Physical health	62 (23.0%)	27 (10.0%)	19 (7.1%)	16 (5.9%)
4. Place to live in	56 (20.8%)	18 (6.7%)	22 (8.2%)	16 (5.9%)
5. Being displaced from home	50 (18.6%)	19 (7.1%)	19 (7.1%)	12 (4.5%)
6. Separation from family members	45 (16.7%)	15 (5.6%)	12 (4.5%)	18 (6.7%)
7. Clothes, shoes, bedding or blankets	44 (16.4%)	5 (1.9%)	16 (5.9%)	23 (8.6%)
8. Alcohol or drug use in your community	39 (14.5%)	10 (3.7%)	13 (4.8%)	16 (5.9%)
9. Care for people in your community who are on their own	21 (7.8%)	7 (2.6%)	6 (2.2%)	8 (3.0%)
9. Health care	21 (7.8%)	6 (2.2%)	9 (3.3%)	6 (2.2%)
9. Distress	21 (7.8%)	6 (2.2%)	9 (3.3%)	6 (2.2%)
12. Toilets	19 (7.1%)	5 (1.9%)	8 (3.0%)	6 (2.2%)
13. Too much free time	18 (6.7%)	7 (2.6%)	4 (1.5%)	7 (2.6%)
13. Mental illness in your community	18 (6.7%)	5 (1.9%)	6 (2.2%)	7 (2.6%)
15. Care for family members	17 (6.3%)	5 (1.9%)	8 (3.0%)	4 (1.5%)
16. Education for your children	16 (5.9%)	4 (1.5%)	11 (4.1%)	1 (0.4%)
16. Safety or protection from violence for women in your community	16 (5.9%)	3 (1.1%)	7 (2.6%)	6 (2.2%)
18. Keeping clean	11 (4.1%)	1 (0.4%)	4 (1.5%)	6 (2.2%)
19. Moving between places	10 (3.7%)	2 (0.7%)	5 (1.9%)	3 (1.1%)

Appendix 3
Example HESPER Report

20.	Safety	9 (3.3%)	5 (1.9%)	2 (0.7%)	2 (0.7%)
21.	The way aid is provided	8 (3.0%)	4 (1.5%)	2 (0.7%)	2 (0.7%)
21.	Law and justice in your community	8 (3.0%)	2 (0.7%)	3 (1.1%)	3 (1.1%)
23.	Drinking water	7 (2.6%)	4 (1.5%)	1 (0.4%)	2 (0.7%)
24.	Respect	3 (1.1%)	1 (0.4%)	0	2 (0.7%)
25.	Support from others	2 (0.7%)	0	0	2 (0.7%)
26.	Information	0	0	0	0

Figure 2: Proportion with which each of the HESPER Scale's problem areas was given a priority rating by participants (i.e. was rated as one of participants' three most serious problems). Only the 12 HESPER problem areas which received the most priority ratings are listed. The remaining 14 problem areas are grouped together under the 'Other' category.

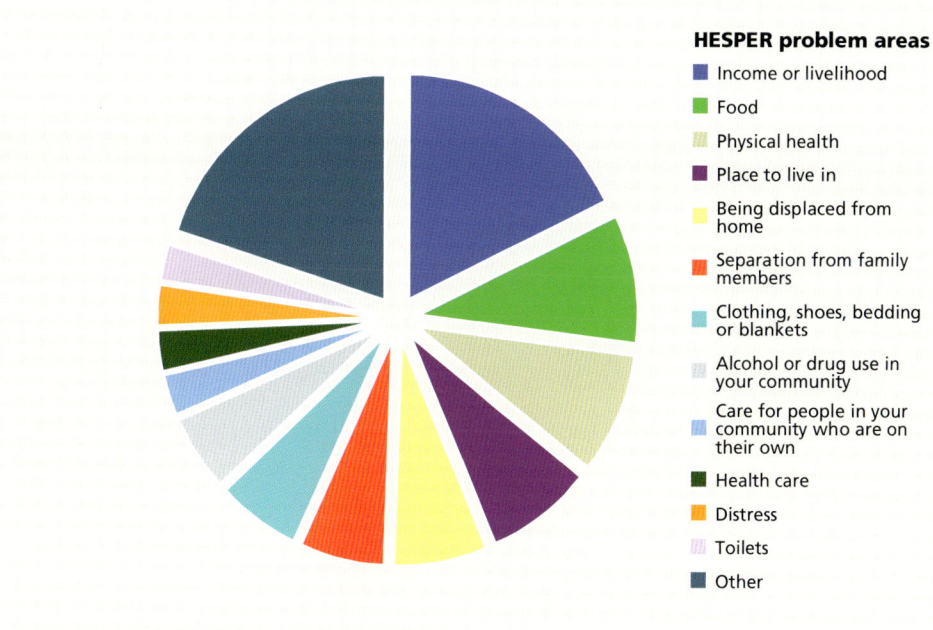

Table 3: Number of participants (% in brackets) who rated each of the HESPER Scale's problem areas as serious problem, no serious problem or did not answer (i.e. not known, not applicable, or answer declined) (n=269). Areas are ranked and listed in descending order of serious problem ratings.

HESPER item	Serious problem	No serious problem	No answer
1. Income or livelihood	202 (75.1%)	67 (24.9%)	0
2. Food	156 (58.0%)	113 (42.0%)	0
3. Being displaced from home	140 (52.0%)	121 (45.0%)	8 (3.0%)
4. Clothes, shoes, bedding or blankets	132 (49.1%)	137 (50.9%)	0
5. Place to live in	120 (44.6%)	149 (55.4%)	0
6. Separation from family members	113 (42.0%)	156 (58.0%)	0
7. Alcohol or drug use in your community	111 (41.3%)	156 (58.0%)	2 (0.7%)
8. Physical health	107 (39.8%)	162 (60.2%)	0
9. Care for people in your community who are on their own	96 (35.7%)	170 (63.2%)	3 (1.1%)
10. Distress	93 (34.6%)	176 (65.4%)	0
11. Too much free time	91 (33.8%)	178 (66.2%)	0
12. Keeping clean	84 (31.2%)	185 (68.8%)	0
13. Care for family members	75 (28.0%)	193 (72.0%)	0
14. Toilets	75 (27.9%)	194 (72.1%)	0
15. Moving between places	70 (26.0%)	199 (74.0%)	0
16. Safety or protection from violence for women in your community	69 (25.7%)	193 (71.7%)	7 (2.6%)
17. Law and justice in your community	67 (24.9%)	192 (71.4%)	10 (3.7%)
18. Health care	67 (24.9%)	201 (74.7%)	1 (0.4%)
19. Mental illness in your community	63 (23.4%)	203 (75.5%)	3 (1.1%)

Appendix 3
Example HESPER Report

20. The way aid is provided	52 (19.3%)	217 (80.7%)	0
21. Safety	45 (16.7%)	224 (83.3%)	0
22. Information	42 (15.6%)	226 (84.0%)	1 (0.4%)
23. Education for your children	36 (13.4%)	201 (74.7%)	32 (11.9%)
24. Respect	32 (11.9%)	237 (88.1%)	0
25. Support from others	29 (10.8%)	240 (89.2%)	0
26. Drinking water	18 (6.7%)	251 (93.3%)	0

Numbers do not always add up to total number of participants due to missing data.

Sri Lanka, following the Indian Ocean tsunami, 2005, © WHO

Appendix 4 - Sampling Guide

SIMPLE RANDOM SAMPLING

Simple random sampling is the most basic and straightforward type of probabilistic sampling. It involves selecting each sampling unit randomly and independently from a list of all sampling units.

Step-by-step guide to simple random sampling
- Obtain a list of all sampling units, i.e. all members of the target population or all households.
- Number each sampling unit on this list.
- Then select sampling units into the study by randomly choosing numbers (e.g. by using a random number table).
- Continue until you have reached your required sample size.
- *If households are used as sampling units:* Randomly select one individual in each chosen household into the study, for instance by using a Kish Table (41), see Appendix 5; also see step 3 under cluster sampling below).

SYSTEMATIC RANDOM SAMPLING

Systematic random sampling is similar to simple random sampling, in that each sampling unit is chosen randomly and independently from a list of all sampling units. However, the method by which sampling units are selected into the study is different.

Step-by-step guide to systematic random sampling
- Obtain a list of all sampling units, i.e. all members of the target population or all households.
- Number each sampling unit on this list.
- Calculate a sampling interval by dividing the number of sampling units by the sample size.
- Randomly select a number between 1 and the number of the sampling interval (e.g. by using a random number table). This is the first sampling unit selected into the study.
- Select each new sampling unit by adding the number of the sampling interval to the previous number.
- Continue until you have reached your required sample size.
- *If households are used as sampling units:* Randomly select one individual in each chosen household into the study, for instance by using a Kish Table (41), see Appendix 5; also see step 3 under cluster sampling below).

CLUSTER SAMPLING

Cluster sampling essentially involves selecting smaller geographical areas (or clusters) from within the target population, and then using simple or systematic random sampling methods within these smaller areas. For this, the target population is first divided according to clusters, such as different areas in a country, different towns or villages in a country or region, different areas of a town, different camps etc. Individuals are then selected into the study by:
1. Randomly selecting a specified number of clusters.
2. Randomly selecting a specified number of households within these chosen clusters.
3. Randomly selecting an individual as study participant from within the chosen households.

The advantages of cluster sampling over simple or systematic random sampling are (40):
- It does not require a complete list of all members of the target population or all households.
- It is cheaper (as individuals selected into the study live more closely to one other).

However, the disadvantages of cluster sampling are that (40):
- It leads to less precise estimates.
- It complicates the statistical analyses.
- It requires larger sample sizes. This is due to the design effect, which means that individuals living in close proximity to each other are more likely to have the same, or similar, characteristics than those not living closely together (i.e. outcomes tend to cluster within populations). The higher the clustering of an outcome in the population, the higher the design effect and the larger the sample size needs to be.

Step-by-step guide to cluster sampling

Step 1 – Selecting clusters
- Obtain or construct a list of all clusters (e.g. towns within a country, camps in a given area), together with their population sizes (if known).
- Decide on the number of clusters to include in the study. It is common in epidemiological surveys to choose 30 clusters, which is often sufficient.
- Calculate your sampling interval by dividing the total population size (of all clusters combined) by the number of clusters that are being included in the study.
- Randomly select a number between 1 and the sampling interval (e.g. by using a random number table).
- Use this random number as start cluster; choose your second cluster by adding the sampling interval and selecting the next cluster accordingly.
- Continue with this until you have reached your required number of clusters.

Step 2 – Selecting households within clusters

Randomly select households within chosen clusters through one of these methods:

- Simple or systematic random sampling of households: Obtain a complete list of all households in the cluster, for example by asking a community leader, obtaining a map of the cluster (e.g. by using GoogleEarth), or by drawing out a map yourself (if clusters are small). Then give each household a number, and randomly select the required number of households by using simple or systematic random sampling techniques (see page 80 above). If there are more than 100 to 200 households in the cluster, divide the cluster into sub-sections, and then list and randomly select households from a randomly selected sub-section.

- Segmentation method: As in the method above, obtain or draw a map of all households in selected clusters. Then divide each cluster into segments of approximately equal size, choose one of these segments from within each cluster at random, and select all households within each of these segments into the study. The size of segments (i.e. the number of households in each segment) should correspond to the number of households required per cluster.

- Random-walk method: As this method is the most prone to bias, it should only be used where the other two methods described above are not feasible. Where there is a map of the cluster, a starting point may be chosen by listing a few different possible starting points on the map at easily identifiable locations, and then randomly selecting one of these. Where there is no map of the cluster, start at the centre of the cluster. Then choose a random walking direction, for example by spinning a pen or a bottle. Walk in a straight line in the selected direction until you reach the edge of the cluster, counting the number of houses in that line. Randomly select a number between 1 and the number of houses in the line (e.g. by using a random number table); this is the first house. Select the next closest house to this house, then the next closest house to that etc (these do not necessarily have to lie in the initial line that was used to select the first house), until you have completed the number of houses required in that cluster.

Step 3 – Selecting individuals within chosen households

- List all individuals within selected households who are eligible for your study (e.g. all members of the household who are 18 years of age or older). Include even those who are absent at the time of your visit, but who usually live there. You can do this by asking the person you have approached to tell you who lives there.

- Randomly select an individual from this list. One way in which you may do this is to use a Kish Table (41) (see Appendix 5).

Appendix 5 - Kish Table

Kish Table

People in the household who are eligible for the study (oldest listed first, youngest last):	Participant number ending in:									
	1	2	3	4	5	6	7	8	9	0
1	1	1	1	1	1	1	1	1	1	1
2	2	1	2	1	2	1	2	1	2	1
3	3	2	1	3	2	1	3	2	1	3
4	4	3	2	1	4	3	2	1	4	3
5	5	4	3	2	1	5	4	3	2	1
6	6	5	4	3	2	1	6	5	4	3

USER GUIDE

- List every eligible respondent in the household, for example every person over 18 years of age. List them in order of their age, with the oldest person listed first, and the youngest last. Include even those people who are absent at the time of your visit but who usually live there. You can do this by asking the person you have approached to tell you who lives there.
- Circle the number to the left of the last person on the list (in the left column).
- In the top row of the table, find the number corresponding to the last digit of the pre-assigned participant number, and circle it.
- Circle the number in the box at which the chosen row and column cross. You should interview the person on your list who corresponds to this number.

Example

- We are imagining that there are three eligible people living in the household. Write down their names in the left-hand column, and then circle the number 3 in the column to the left (i.e. the number of eligible people in the household).
- We are imagining that the pre-assigned participant number is 68. Circle the number 8 in the row at the top of the table (i.e. the last digit of the participant number).
- Find the number in the box at which the chosen row and column cross; in this case it is the number 2. The person you should interview would therefore be the second person on the list in the left-hand column.

Appendix 6 - Performing Sample Size Calculations

In order to perform a sample size calculation, you will need to decide:
- The likely prevalence of your outcome (which in HESPER surveys is 'perceived needs'). To estimate this, you could look at previous similar surveys of needs, or previous qualitative interviews. As it may often be difficult to estimate the prevalence of perceived needs for each of the HESPER Scale's 26 items, it may often be appropriate to assume a prevalence of 50%, as this will give you the largest (and therefore most conservative) sample size estimate (i.e. you will be erring on the side of caution).
- How precisely your outcome should be measured (38). For this, you will need to determine both the required level of precision in your study, and also the highest acceptable level of error. These will both depend on the reasons for the study and on the resources available (38). For HESPER surveys, a level of precision of 10%, and a risk of error of 5%, should usually be adequate.

The formula to calculate a required sample size is (as long as the target population includes at least a few thousand people) (40):

$$n = \frac{t^2 \times p \times q}{d^2}$$

where:

n   is the required sample size

t   is a value related to the risk of error (where the risk of error is 5%, a figure of 1.96 should be used for this)

p   is the expected prevalence (reported as a fraction of 1, e.g. 0.5)

q   is the expected non-prevalence (i.e. 1-p)

d   is the level of precision (also reported as a fraction of 1, e.g. 0.1)

As an example, where the expected prevalence is 50% (or 0.5), the level of precision is 10% (or 0.1), and the risk of error is 5% (as may commonly be appropriate for HESPER surveys), the required sample size would be as follows:

$$n = \frac{1.96^2 \times 0.5 \times 0.5}{0.1^2} = \frac{3.8416 \times 0.25}{0.01} = 96.04$$

The required sample size (i.e. the minimum number of people needed to participate in the study) would be 96. The level of precision of 10% would, in this case, imply a likely range for the true value of between 40% and 60%, and the risk of error would imply that there was a 5% chance that the true value would lie outside this range.

However, the calculated sample size would then also need to be adjusted according to the following factors:
- Margin for non-response (for all studies).
- Design effect (only where cluster sampling methods have been used).

This is done as follows:
- Calculate the minimum number of people needed for the study by using the formula above:
$$n = \frac{t^2 \times p \times q}{d^2}$$
- If you are using cluster sampling methods, multiply this number (n) by the design effect. Skip this step if you are using simple or systematic random sampling methods.
- Divide the obtained number by the expected response rate to account for non-response.
- The resulting number is the number of people you need to select into your study.

Sample size adjustment for non-response

Not all respondents who are invited to participate in a survey will take part. The most common reasons are that respondents are not at home, or that respondents decline to participate (38).

Sample size requirements always need to be adjusted to account for this non-response. You may establish the likely non-response rate by looking at similar surveys that have been conducted in the target population previously. Non-response rates may vary widely depending on the setting, target population and sampling method. Generally, although a non-response rate of 30% is considered adequate, often non-response rates in humanitarian settings may be less than 10%.

The formula to account for non-response is:

$n_{\text{accounting for non-response}} = n_{\text{not accounting for non-response}}$ / expected response (i.e. 1 − expected non-response)

In our example above, if the non-response rate was expected to be 10%, the following calculation would be performed:

$n_{\text{accounting for non-response}} = 96 / 0.9 = 106.67$

The adjusted sample size (of people to approach for the study) would be 107.

Sample size adjustment for cluster sampling

In studies where a cluster sampling method is used, the required sample size will need to be multiplied by the design effect. Required sample sizes are therefore higher in studies using cluster sampling methods than in those employing simple or systematic random sampling techniques. For instance, for a survey employing cluster sampling methods in which the design effect is 2, the required sample size would be twice that of one using simple or systematic random sampling. In practice, though design effects vary according to type of outcome and location, a design effect of 2 is commonly used.

The formula to calculate the required sample size in studies where cluster sampling has been employed is:

$$n_{\text{accounting for design effect and non-response}} = \frac{(t^2 \times p \times q)}{d^2} \times \text{design effect} / \text{expected response}$$

In our example above, with a design effect of 2 and a margin of non-response of 10%, the sample size calculation would be:

$$n_{\text{accounting for design effect and non-response}} = ((1.96^2 \times \frac{0.5 \times 0.5}{0.1^2}) \times 2) / 0.9 = 96.04 \times 2 / 0.9 = 213.42$$

The required sample size (of people to approach for the study) would be 213.

Appendix 7 - Calculating Confidence Intervals

CONFIDENCE INTERVALS FOR PREVALENCE

The formula to calculate 95% confidence intervals for prevalence is as follows:

95% CI = prevalence +/- 1.96 x standard error (SE) of the prevalence

The standard error (SE) for prevalence (P) is calculated by:

$SE = \sqrt{P(100-P)/n}$

For example, if in a sample of 200 respondents 25% of respondents rated the HESPER item 'Drinking water' as a serious problem, the 95% confidence intervals would be calculated as follows:

$SE = \sqrt{P(100-P)/n} = \sqrt{25(100-25)/200} = \sqrt{25 \times 75/200} = \sqrt{1875/200} = 3.06$

95%CI = 25 − 1.96 x 3.06 = 19.0%
95%CI = 25 + 1.96 x 3.06 = 31.0%

95%CI = 19.0% − 31.0%

This means that if we repeated the same study 100 times under the same conditions but with a different sample, in 95 of these samples the proportion of respondents who rated the HESPER item 'Drinking water' as serious problem would lie between 19% and 31%. The true value in the target population would be likely to lie within this range (with 95% certainty), with the most likely true value being 25%.

Appendix 7 continued on next page

CONFIDENCE INTERVALS FOR MEAN

The formula to calculate 95% confidence intervals for a mean is as follows:

95% CI = mean +/- 1.96 x standard error (SE) of the mean

The standard error (SE) for a mean is calculated by:
standard deviation (SD) / $\sqrt{n}$.

For example, if in a sample of 100 respondents on average (mean figure) respondents rated 12.0 (SD 2.5) of the HESPER items as serious problem, 95% confidence intervals for this mean would be calculated as follows:

SE = 2.5 / $\sqrt{100}$ = 2.5 / 10 = 0.25

95%CI = 12.0 − 1.96 x 0.25 = 11.51
95%CI = 12.0 + 1.96 x 0.25 = 12.49

95%CI = 11.51 − 12.49

This means that if we repeated the same study 100 times under the same conditions but with a different sample, in 95 of these samples the mean number of 'serious problem' ratings by respondents would lie between 11.51 and 12.49. The true value in the target population would be likely to lie within this range (with 95% certainty), with the most likely true value being 12.0.

Appendix 8 - Example Participant Information Sheet / Consent Form

EXAMPLE PARTICIPANT INFORMATION SHEET / CONSENT FORM

Hello, my name is …We are inviting you to take part in an assessment by *insert your agency / organisation*. We are conducting an assessment to find out about the serious problems that people have when they have experienced a conflict or another disaster. We hope that by better understanding what people like you see as their serious problems, more people will get the help they really want.

I would like to assure you that participation in this assessment is voluntary. I can also assure you that all the information we receive will be completely confidential, so it will not be possible for anybody outside our team to link any of the information we collect to you.

If you decide to take part, we would invite you to meet with the interviewer on one occasion. The interview would take about 15 to 30 minutes of your time and we would ask you questions about the serious problems you may currently be experiencing.

You can either start the assessment now, or you can let us know within the next few days whether you would like to take part. If you decide to take part, you have the right to decline to answer any question I ask you. Please just let me know and I will move to the next question. You may also stop the interview at any time if you wish and without having to give a reason. Unfortunately we will not be able to offer you or your family any compensation or other benefits if you decide to take part.

If you have any questions now or in the future you can contact *insert your organisation's address and telephone number* for further advice and information.

Thank you for your time.

Do you have any questions?

Do you agree to be in this assessment?   **Yes**   **No**

_____   _____
Signature   Date

either to be signed by participant
(where written consent is taken), or
by interviewer as witness to participants'
consent (where verbal consent is taken)

Displacement camp in Port-au-Prince, Haiti, 2010.
© Maya Semrau